Michael Schinhammer

Development and characterization of biodegradable Fe-based alloys

Michael Schinhammer

Development and characterization of biodegradable Fe-based alloys

for temporary medical implant applications

Südwestdeutscher Verlag für Hochschulschriften

Impressum / Imprint
Bibliografische Information der Deutschen Nationalbibliothek: Die Deutsche Nationalbibliothek verzeichnet diese Publikation in der Deutschen Nationalbibliografie; detaillierte bibliografische Daten sind im Internet über http://dnb.d-nb.de abrufbar.
Alle in diesem Buch genannten Marken und Produktnamen unterliegen warenzeichen-, marken- oder patentrechtlichem Schutz bzw. sind Warenzeichen oder eingetragene Warenzeichen der jeweiligen Inhaber. Die Wiedergabe von Marken, Produktnamen, Gebrauchsnamen, Handelsnamen, Warenbezeichnungen u.s.w. in diesem Werk berechtigt auch ohne besondere Kennzeichnung nicht zu der Annahme, dass solche Namen im Sinne der Warenzeichen- und Markenschutzgesetzgebung als frei zu betrachten wären und daher von jedermann benutzt werden dürften.

Bibliographic information published by the Deutsche Nationalbibliothek: The Deutsche Nationalbibliothek lists this publication in the Deutsche Nationalbibliografie; detailed bibliographic data are available in the Internet at http://dnb.d-nb.de.
Any brand names and product names mentioned in this book are subject to trademark, brand or patent protection and are trademarks or registered trademarks of their respective holders. The use of brand names, product names, common names, trade names, product descriptions etc. even without a particular marking in this works is in no way to be construed to mean that such names may be regarded as unrestricted in respect of trademark and brand protection legislation and could thus be used by anyone.

Coverbild / Cover image: www.ingimage.com

Verlag / Publisher:
Südwestdeutscher Verlag für Hochschulschriften
ist ein Imprint der / is a trademark of
AV Akademikerverlag GmbH & Co. KG
Heinrich-Böcking-Str. 6-8, 66121 Saarbrücken, Deutschland / Germany
Email: info@svh-verlag.de

Herstellung: siehe letzte Seite /
Printed at: see last page
ISBN: 978-3-8381-2833-7

Zugl. / Approved by: Zurich, ETH Zurich, Diss., 2012

Copyright © 2013 AV Akademikerverlag GmbH & Co. KG
Alle Rechte vorbehalten. / All rights reserved. Saarbrücken 2013

für Fabienne und meine Eltern

Acknowledgements

A project such as a Ph.D. is not the work of a single person alone but the achievement of all the persons involved. Here I would like to thank everyone who contributed. First and foremost I would like to express my sincere gratitude to Prof. Peter Uggowitzer, who supervised and guided this project. He motivated me to do a Ph.D. on the fascinating topic of biodegradable metals, and his staunch support was invaluable to the project. I appreciate his profound knowledge, his spirit and the enriching discussions on both scientific and other topics.

I would like to thank Prof. Jörg Löffler for making it possible to carry out this project in his group and for being co-referee of this thesis. I appreciate the confidence he has in his group and the freedom he provides us. I also gratefully acknowledge Prof. Annelie Weinberg and Prof. Ewald Werner for co-refereeing this work.

I thank Dr. Anja Hänzi sincerely for her great support during my Master's thesis and at the beginning of this project. She is responsible for many valuable ideas and I learned a lot about scientific work from her. Sincere thanks also go to my current (Shankar Kumar Jha, Frank Moszner, Jérôme Zemp) and former (Drs. Petra Gunde, Konrad Papis, Dirk Uhlenhaut, Bruno Zberg) office mates and fellow doctoral students (especially David Klaumünzer and Reto Giannini) who all contributed to a fruitful working atmosphere. I am also thankful for the help of all other group members and colleagues. Much knowledge and support came from Erwin Fischer, who helped with casting; Christian Wegmann, whose intuitive talent for practical things made my work easier; Markus Müller and Christian Roth for sample preparation; Katherine Hahn Halbheer for correcting my manuscripts; and of course Annika Reisacher, Beatrice Wegmann, Joe Hecht, Edi Schaller, and Carlo Bernasconi.

I am grateful to Dr. Alla Sologubenko for sharing her extensive knowledge of microstructural characteristics and electron microscopy. It helped me a lot during sample analysis and interpretation. I also gratefully acknowledge the support of Dr. Elisabeth Müller, Dr. Fabian Gramm and Dr. Karsten Kunze from the Electron Microscopy Center. They contributed to this project with TEM training support regarding TEM and EBSD analysis. I would also like to thank Dr. Florian Dalla Torre for his initial TEM studies, Dr. Isabel Gerber for performing the cell tests, and Dr. Patrik Schmutz for discussions on corrosion and his enthusiasm for this branch of science. I would also like to acknowledge Stefan Beck and Thomas Imwinkelried for their collaboration.

During the course of this project I was able to supervise a number of students who were interested in this project. They contributed substantially to my thesis and I always enjoyed working with them: Fabian Fischer, Simon Gruber, Frank Moszner, Lea Nowack, Christina Pecnik, Felix Rechberger, Cédric Sax, Patrick Steiger, Deborah Solenthaler, Andi Wyss, Carmen Ziegler, and Rahel Zoller.

Last but not least I would like to express my gratitude to my friends and family, who always supported me during this task even though they sometimes had to endure lengthy explanations of microstructure, degradation properties and so forth. They always listened with a smile.

I gratefully acknowledge the financial and scientific support received within the framework of the project "Biocompatible Materials and Applications" initiated by the Austrian Institute of Technology GmbH (AIT), and support from the Staub/Kaiser Foundation, Switzerland.

Contents

Abstract ix
Zusammenfassung xii

1 Introduction   **1**
  1.1 General introduction 2
  1.2 Potential applications for degradable metals 6
      1.2.1 Cardiovascular stents 6
      1.2.2 Elastic stable intramedullary nailing 9
      1.2.3 Further applications 10
  1.3 Iron as biodegradable material 11
  1.4 Design strategy for Fe-based degradable implant materials 12
      1.4.1 Electrochemical considerations 12
      1.4.2 Formation of precipitates 14
      1.4.3 In vivo corrosion 16
      1.4.4 Biocompatibility considerations 18
      1.4.5 Aim of the project and outline of the thesis 20
  References 21

2 Design strategy   **29**
  2.1 Introduction 30
  2.2 Design strategy 32
  2.3 Materials and methods 36
      2.3.1 Sample preparation and heat treatments 36
      2.3.2 Microstructure and characterization 38
      2.3.3 Immersion testing and impedance spectroscopy 38

	2.3.4	Mechanical testing	39
2.4	Results		40
	2.4.1	Microstructure	40
	2.4.2	Immersion testing and impedance spectroscopy	43
	2.4.3	Mechanical properties	44
2.5	Discussion		45
	2.5.1	Microstructure	45
	2.5.2	Influence of the microstructure on the degradation performance	48
	2.5.3	Influence of the microstructure on the mechanical performance	49
	2.5.4	Efficiency and potential of the design strategy	50
2.6	Conclusions		51
	References		52

3 Microstructure and mechanical performance — **55**

3.1	Introduction		56
3.2	Experimental		58
	3.2.1	Alloy preparation and thermo-mechanical treatments	58
	3.2.2	Microstructure characterization	60
3.3	Results		61
	3.3.1	Influence of the annealing temperature on recrystallization behavior	61
	3.3.2	Influence of the degree of cold working on recrystallization behavior	66
	3.3.3	Thermo-mechanical optimization	66
3.4	Discussion		71
	3.4.1	Recrystallization behavior	71
	3.4.2	Thermo-mechanical optimization	76
3.5	Conclusions		79
	References		80

4 Degradation properties I - Immersion testing — 85

- 4.1 Introduction 86
- 4.2 Materials and Methods 88
  - 4.2.1 Materials 88
  - 4.2.2 Immersion testing setup 89
- 4.3 Results 92
  - 4.3.1 Immersion testing of pure iron 92
  - 4.3.2 Immersion testing of WZ21 92
- 4.4 Discussion 94
  - 4.4.1 The use of CO_2 as a pH buffer 95
  - 4.4.2 Influence of the testing conditions on degradation rates . . 96
  - 4.4.3 Correlation with in vivo results 99
- 4.5 Conclusions 100
- References 101

5 Degradation properties II - Degradation performance of Fe–Mn–C(–Pd) alloys — 107

- 5.1 Introduction 108
- 5.2 Materials and Methods 110
  - 5.2.1 Materials 110
  - 5.2.2 Methods 111
- 5.3 Results 115
  - 5.3.1 Microstructure characterization 115
  - 5.3.2 Immersion testing in SBF 116
  - 5.3.3 Cross-sections of immersed samples 118
  - 5.3.4 Electrochemical impedance spectroscopy in SBF 120
  - 5.3.5 Immersion testing in H_2SO_4 124
- 5.4 Discussion 125
  - 5.4.1 Degradation behavior in SBF 125
  - 5.4.2 Degradation mechanism in SBF 128
  - 5.4.3 Influence of the precipitates on degradation behavior . . . 133
  - 5.4.4 Influence of degradation products 134

	5.4.5	Influence of the testing conditions on degradation behavior	135
5.5	Conclusions		135
References			137

6 Biocompatibility aspects — 143
- 6.1 Introduction . . . 144
- 6.2 Materials and methods . . . 146
  - 6.2.1 Materials . . . 146
  - 6.2.2 Cytocompatibility studies . . . 146
  - 6.2.3 Ion release . . . 148
- 6.3 Results . . . 149
  - 6.3.1 Pilot cytocompatibility studies . . . 149
  - 6.3.2 Main cytocompatibility studies . . . 149
  - 6.3.3 Dose-response curves . . . 150
  - 6.3.4 Ion release . . . 151
  - 6.3.5 Analysis of sample surface after eluate preparation . . . 152
- 6.4 Discussion . . . 153
  - 6.4.1 The alloying elements' influence on metabolic processes and their cytocompatibility . . . 154
  - 6.4.2 Cytocompatibility of TWIP-steels . . . 159
  - 6.4.3 Remarks on experimental setup . . . 161
- 6.5 Conclusions . . . 161
- References . . . 162

7 Summary and Outlook — 167
- 7.1 Summary . . . 168
- 7.2 Outlook . . . 171

Abstract

Iron and its alloys are currently evaluated for use as biodegradable materials for temporary medical implants. The concept of a metallic implant that degrades within the human body has become increasingly accepted in recent years, even though it diverges from another current paradigm which holds that metallic implants should be inert and corrosion-resistant. In a number of applications patients would in fact benefit from degradable implant materials: metallic implants such as osteosynthesis screws, plates and nails usually require removal after the tissue has healed, for example. If the material degraded in the body while the tissue recovered implant removal would become obsolete. In addition, permanent cardiovascular stents cause undesired long-term risks and side effects which might potentially be reduced if they were degradable.

Fe is an interesting candidate as a degradable implant material. However, pure Fe degrades at too low a rate in vivo, and also has poor mechanical properties. To address these points, this work develops a design strategy which takes into account electrochemical, microstructural and toxicological considerations. The aim is to find alloys whose performance is suitable for temporary implant applications, in terms of both increased degradation rate and appropriate strength and ductility.

The design strategy deploys two approaches. One is to add an electrochemically less noble alloying element, in order to make the entire matrix more susceptible to corrosion. The other is to add a small amount of a noble element, with the intention of generating noble precipitates which cause microgalvanic corrosion. Mn and Pd have proved to be suitable alloying elements in this context.

The design strategy focuses primarily on modifying degradation properties, but the alloys' mechanical performance is also very important. The austenitic high-Mn-content alloys developed in the study possess an impressive combination of high ductility, simultaneously high strength levels and a pronounced strain hardening response. They can also be adjusted over a wide range, depending on the requirements of the application. A comparison with materials commonly deployed in permanent implants revealed that the new alloys actually perform better than the standard ones.

Pd-rich precipitates formed in the cold-worked state were found to considerably retard recrystallization during annealing treatment. Electron microscopy was used to investigate these precipitates and their interaction with dislocations. It identified both reduced dislocation mobility caused by a solute drag effect, produced by enrichment of dislocation cores with Pd, and grain boundary pinning (Zener drag) as the mechanisms which hinder recrystallization. The interplay between precipitation and recrystallization was incorporated into a model, which served as a basis for developing the thermomechanical treatments required to achieve microstructural characteristics and mechanical properties suitable for degradable implants.

Degradation properties were evaluated by means of immersion testing and electrochemical impedance spectroscopy in physiological media. To more accurately mimic the in vivo conditions, pH buffers were avoided. Here a new means of pH control was incorporated into the experimental setup: gaseous CO_2 was used to control the pH value in a narrow range around the physiological value of 7.40. Both measurement techniques revealed that Fe–Mn–C–Pd alloys feature an increased degradation rate compared to pure Fe. Electrochemical measurements turned out to be a sensitive tool for investigating degradation behavior. Characteristic values can be deduced by using an equivalent circuit to fit the impedance spectra. One of the most important values is polarization resistance, which is measures corrosion tendency. Electrochemical measurements also provide information on the evolution of degradation product layers.

In vitro cytocompatibility studies were performed by means of indirect cell tests using eluates (extracts) from the Fe-based alloys. Experiments with human umbilical vein endothelial cells revealed acceptable cytocompatibility, based on the alloys' eluates. A chemical analysis of the eluates revealed that it was mainly Mn which dissolved, and that Fe precipitated as insoluble degradation products. The results were discussed with reference to dose-response curves of the main alloying elements Fe and Mn. An important finding for future alloy development was that a high Mn release from implants may limit an alloy's biocompatibility.

The overall performance of the alloys developed and characterized in this study is very promising. The mechanical performance of the austenitic high-Mn-content alloys is especially interesting in terms of their potential application as degradable implant material. In addition, the insights gained may also benefit other efforts to optimize the performance of Fe-based alloys.

Zusammenfassung

Eisen und Eisenbasis-Legierungen werden zurzeit hinsichtlich ihrer Eignung als degradierbare Implantatwerkstoffe untersucht. Das Konzept, dass ein metallisches Implantat im menschlichen Körper degradiert, hat sich in den letzten Jahren zunehmend verbreitet und wird mittlerweile allgemein akzeptiert. Dies obwohl es von der weit verbreiteten Meinung abweicht, dass metallische Implantate inert und korrosionsbeständig sein sollen. In vielen Anwendungen könnten degradierbare Implantate zum Wohl der Patienten eingesetzt werden: Metallische Implantate wie z.B. Osteosynthese-Schrauben, Platten und Nägel müssen nach der Gewebeheilung wieder entfernt werden. Wenn das Material im Körper degradieren würde, während das Gewebe heilt, könnte auf einen zweiten Eingriff zur Entfernung des Implantats verzichtet werden. Weiter wird erwartet, dass die Langzeit Risiken und Nebenwirkungen, welche bei permanenten kardiovaskulären Stents auftreten, durch den Einsatz von degradierbaren Materialien verringert werden können.

Fe ist ein interessantes Element für den Einsatz als degradierbares Material. Allerdings degradiert reines Fe zu langsam in vivo und ausserdem sind die mechanischen Eigenschaften ungenügend. Um diese Punkte anzugehen, wird in dieser Arbeit eine Design Strategie vorgestellt, welche auf elektrochemischen, mikrostrukturellen und toxikologischen Überlegungen beruht. Das Ziel ist es, Legierungen zu entwickeln, welche die Anforderungen hinsichtlich einer erhöhten Abbaurate und angemessener Festigkeit und Duktilität erfüllen.

Die Design Strategie verfolgt zwei Ansätze: Der erste ist es, ein elektrochemisch gesehen unedleres Legierungselement zu verwenden, um die Matrix korrosionsanfälliger zu machen. Der zweite beruht auf der Zugabe einer kleinen Menge eines edleren Legierungselements, mit der Absicht edle Ausscheidungen zu erzeugen. Diese führen dann zu mikrogalvanischer Korrosion. Mn und Pd sind geeignete Legierungselemente um diese Ansätze zu verfolgen.

Die Design Strategie legt einerseits grossen Wert auf die Modifikation der Degradationseigenschaften, aber andererseits dürfen auch die mechanischen Eigenschaften nicht ausser Acht gelassen werden. Die austenitischen hoch-Mn-haltigen Legierungen, die im Rahmen dieser Arbeit entwickelt wurden, besitzen eine beeindruckende Kombination aus hoher Duktilität, hoher Festigkeit und ausgeprägter Verfestigung. Diese Kennwerte lassen sich über einen weiten Bereich einstellen, abhängig von den Anforderungen der Anwendung. Ein Vergleich mit den bislang üblichen Materialien zeigte, dass diese von den neu entwickelten Legierungen überflügelt werden.

Die Pd-reichen Ausscheidungen, welche in einer kaltverformten Probe während einer Wärmebehandlung gebildet werden, behindern deutlich die gleichzeitig auftretende Rekristallisation. Mittels Elektronenmikroskopie wurden die Ausscheidungen und deren Wechselwirkung mit Versetzungen untersucht. Zwei Mechanismen, nämlich eine verringerte Mobilität von Versetzungen (Solute Drag-Effekt durch die Anreicherung der Versetzungskerne mit Pd) und eine Behinderung der Korngrenzenwanderung (Zener Drag) wurden identifiziert, welche die Rekristallisation behindern. Der Zusammenhang zwischen Ausscheidungsbildung und Rekristallisation wurde in einem Modell dargestellt, welches als Grundlage für die Entwicklung von angepassten thermomechanischen Prozessen diente. Diese wurden so gewählt, dass für die Anwendung als degradierbare Implantate geeignete Eigenschaften (hinsichtlich Mikrostruktur und mechanischen Eigenschaften) erreicht werden können.

Die Abbaueigenschaften wurden mittels Einlegetests und elektrochemischer Impedanzspektroskopie in physiologischem Medium untersucht. Um die in vivo Bedingung besser abbilden zu können, wurde auf den Einsatz eines pH Puffers verzichtet. Stattdessen wurde mit dem Einsatz von gasförmigem CO_2 zu Pufferung ein neuer Weg eingeschlagen, um den pH um den physiologischen Wert von 7.40 zu stabilisieren. Beide Messmethoden zeigten, dass die neu entwickelten Fe–Mn–C–Pd Legierungen eine höhere Abbaurate als reines Fe besitzen. Elektrochemische Impedanzspektroskopie ist eine empfindliche Messmethode um die Degradationseigenschaften zu untersuchen. Mittels Einsatz eines passenden Ersatzschaltkreises lassen sich charakteristische Kennwerte aus dem Impedanzspektren gewinnen. Einer der wichtigsten Kennwerte ist der Polarisationswiderstand, welcher ein Mass für die Korrosionsbeständigkeit ist. Elektrochemische Messungen können auch dazu verwendet werden, die zeitliche Entwicklung von Degradationsprodukten zu bestimmen.

In vitro Zytokompatibilitätsstudien wurden mittels indirekten Zelltests unter Verwendung von Eluaten (Extrakten) der Fe-Legierungen durchgeführt. Die Experimente mit humanen Nabelschnur-Endothelzellen zeigten eine genügende Zytokompatibilität, basierend auf den Eluaten. Eine chemische Analyse der Eluate zeigte, dass vor allem Mn in Lösung geht, während Fe zu einem überwiegenden Teil in unlöslichen Degradationsprodukten gebunden ist. Die Ergebnisse wurden unter Berücksichtigung von Dosis-Wirkungskurven für Fe und Mn diskutiert. Ein wichtiges Ergebnis für zukünftige Legierungsentwicklung ist die Tatsache, dass eine zu hohe Mn-Freisetzung die Biokompatibilität einer Legierung beschränkt.

Insgesamt wird das Eigenschaftsprofil der im Rahmen dieser Studie entwickelten und charakterisierten Legierungen als vielversprechend eingeschätzt. Die mechanischen Eigenschaften der hoch-Mn-haltigen Legierungen sind besonders gut geeignet für den Einsatz als abbaubare Implantate. Die in dieser Studie gewonnenen Erkenntnisse lassen sich auch auf andere Legierungssysteme übertragen und können dazu beitragen, deren Eigenschaften weiter zu optimieren.

1 Introduction

1 Introduction

1.1 General introduction

Biomaterials are used to make devices that replace a part or a function of the body. In this context a large number of requirements must be taken into account. The biomaterials in question have to carry out their mission in a safe, reliable, economic, and physiologically acceptable manner [1]. An early definition by Bruck [2] states that: "[biomaterials are] materials of synthetic as well as of natural origin in contact with tissue, blood, and biological fluids, and intended for use for prosthetic, diagnostic, therapeutic, and storage applications without adversely affecting the living organism and its components." The successful development of biomaterials hence presupposes broad knowledge and/or collaboration with specialists from relevant areas in medicine, biology and materials science. Fig. 1.1 summarizes some of the knowledge required.

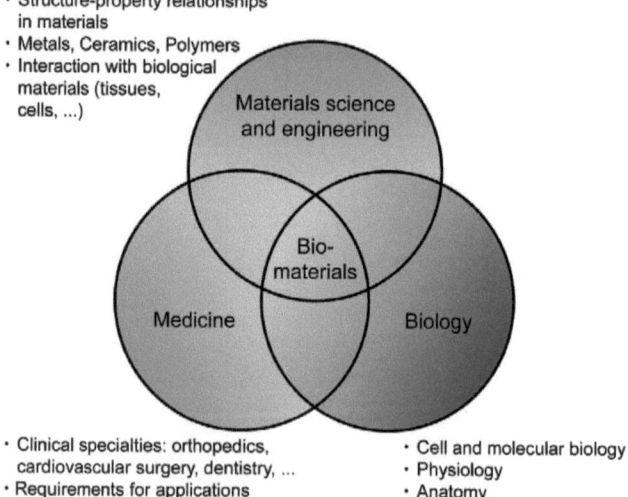

Figure 1.1: Biomaterials are at the interface of materials science, medicine and biology.

The most prominent uses of biomaterials are listed in Table 1.1 [1]. From the table it is evident that even today, most medical implants either bear mechanical loads (e.g., as artificial hip joints) or replace simple physical functions (e.g., as artificial arteries) [3]. Complex chemical functions such as those carried out in the kidneys or the liver, and electrical or electrochemical functions such as those of the brain and sense organs, cannot be performed by biomaterials at this stage [1].

Table 1.1: Overview of uses of biomaterials (from [1]).

Problem area	Examples
Replacement of diseased or damaged part	Artificial hip joint, kidney dialysis machine
Assist in healing	Sutures, bone plates and screws, elastic stable intramedullary nailing
Improve function	Cardiac pacemaker, intraocular lens
Correct functional abnormality	Cardiac pacemaker
Correct cosmetic problem	Augmentation mammoplasty, chin augmentation
Aid to diagnosis	Probes and catheters
Aid to treatment	Catheters, drains

An artificial hip joint made from a high-strength material with an appropriate mechanical design can naturally assume the loads incurred by body movement, but in addition to providing function it must be scrutinized for biological performance [4]. In the European Society for Biomaterials Consensus Conference the term "biocompatibility" was defined as "the ability of a material to perform with an appropriate host response in a specific application" [5]. This definition highlights the importance of acceptance of an artificial implant by the surrounding tissue and by the body as a whole [1]. In particular, truly biocompatible materials are characterized by the absence of unwanted tissue-implant reactions such as abnormal inflammation responses and allergic or immunologic reactions, and they do not cause cancer [1]. However, "biocompatibility" implies not only the mere absence any of unwanted reactions, but also satisfactory performance in the intended biomedical application [4]. The term "biological performance" emphasizes the interaction between living systems and materials. These are, on one hand, the host response (the living system's local and systemic response to the material), and, on the other, the material's response to the living system [4]. Developments and new understanding in materials science have contributed substantially to the success of modern medicine and surgery. Early implants consisted of materials adapted from crafts or industry, and they provoked severe foreign body and inflammation reactions [3]. Most early surgical procedures were also unsuccessful because of infection [1]. Only with the introduction of aseptic surgical techniques by Dr. J. Lister in the 1860s did the situation change and the deployment of biomaterials become feasible. Initially, the most important biomaterials application was in the skeletal system: early metal devices, including wires and pins made from Fe, Au, Ag and Pt, were used to fix bone fractures [1, 6]. Subsequently, the evolution of medical implants was closely linked to the development of new materials. After the introduction of stainless steel and cobalt chromium alloys in the 1930s, reliable (in terms of mechanical performance) and biocompatible (as these alloys are generally inert and do not corrode) bone fixation devices and joint replacements were developed [1].

Nowadays not only metals but all classes of material (metals, ceramics, polymers and composites) are being deployed as biomaterials for various purposes. Table 1.2 lists the various uses and their advantages and disadvantages in the body [1]. Each material is selected and deployed for its specific advantages: hydroxyapatite coatings on dental implants, for example, provide excellent bone ingrowth and consequent stable long-term tissue integration despite poor mechanical strength. Metals are used where their good mechanical performance is a boon, e.g. in load-bearing orthopedic implants and fixation devices such as joint replacements, screws, plates, dental implants, wires and coronary stents [1, 6-10]. Their high strength guarantees good primary stability after surgery, and their fracture toughness, ductility and high fatigue limit are also beneficial in applications.

Traditionally, most metallic implants are made of corrosion-resistant materials such as stainless steel 316L, Ti and its alloys or Co–Cr alloys. They are therefore associated with permanent applications (e.g. joint replacements and stents). Some implants, however, must be removed when they are no longer

Table 1.2: Use of materials in the body (from [1]).

Materials	Advantages	Disadvantages	Examples
Polymers (nylon, silicone rubber, polyester, etc.)	Resilient Easy to fabricate	Not strong Deforms with time (creep) May degrade	Sutures, blood vessels, hip socket, ear, nose, other soft tissues
Metals (Ti and its alloys, Co–Cr alloys, stainless steels, Au, Ag, Pt, etc.)	Strong, tough, ductile	May corrode Dense Difficult to make	Joint replacements, bone plates and screws, dental root implants, pacer and suture wires
Ceramics (aluminum oxide, calcium phosphates including hydroxyapatite)	Very biocompatible Inert Strong in compression	Brittle Not resilient Difficult to make	Dental, femoral head of hip replacement, coating of dental and orthopedic implants
Composites (carbon-carbon, wire or fiber reinforced bone cement)	Strong, tailor-made	Difficult to make	Joint implants, heart valves

needed (e.g., osteosynthesis plates and screws). This requires a second intervention and causes additional pain and discomfort for the patient [11-13]. Permanent implants are also accompanied by long-term risks and side effects. In cardiovascular stents, for example, these include permanent physical irritation, chronic inflammatory reactions, in-stent restenosis, and mismatches in mechanical behavior between stented and non-stented vessel areas [14-16].

Biodegradable materials, on the other hand, degrade in vivo, either via hydrolytic mechanisms (with or without the help of enzymes), resorption or electrochemical reactions [1]. Generally speaking, this can be described as the breakdown of a material mediated by a biological system [4]. This concept is familiar mainly in the context of polymeric materials (e.g. hydrogels or various degradable polymers such as polylactic acid) and ceramics (e.g. calcium phosphates) [1]. Recently, however, a concept of biodegradable metals has been developed which questions the traditional paradigm that metallic biomaterials must be corrosion-resistant [17]. In the following, the significance of this new concept is illustrated in the context of two potential degradable metals applications: cardiovascular stents and elastic stable intramedullary nailing (ESIN).

1.2 Potential applications for degradable metals

1.2.1 Cardiovascular stents

The "Global Burden of Disease" study found ischemic heart disease to be the leading cause of mortality worldwide [18]; it continues to be the biggest killer, especially in the Western world [19, 20]. Ischemia refers to insufficient blood supply to tissues, causing a shortage of the oxygen and nutrition required to maintain cellular metabolism. Blockage of a coronary artery will eventually lead to myocardial damage, heart failure and cardiac death.

The associated (chronic) disease is called atherosclerosis, which is initiated and accompanied by inflammatory processes in the arterial vessel walls [21, 22]. The first observable alteration of the vessel wall is the appearance of "fatty streaks", which are an aggregation of lipids and T lymphocytes in the innermost layer of the artery wall [21, 23]. Many of these lipids (lipoproteins, especially low-density lipoproteins) originate from the plasma and enter the endothelium, where they become oxidized or otherwise modified [21, 23]. They then cause a cascade of processes which eventually generate the fatty streaks. The latter may subsequently develop into more advanced, complex and occlusive lesions, called fibrous plaques [21]. As they grow into the arterial lumen, they increasingly impede blood flow. Moreover, as their size increases there is an increased possibility of plaque breakdown, which is accompanied by the risk of blood clots on the plaque surface [23]. If a clot becomes large enough it can partially or completely block the blood flow through the artery and cause a heart attack. A thrombus may also travel to other organs such as the brain or lung.

Today the standard treatment for coronary artery stenosis (narrowing of the artery) is percutaneous transluminal coronary angioplasty (PTCA), which is employed whenever possible. Because it is a minimally invasive technique it is not only cost-effective but is also preferable in terms of patient comfort (no general anesthesia; short recovery time) and safety. However, for patients with severe stenosis coronary artery bypass surgery may be the better option.

During PTCA a small catheter equipped with an inflatable balloon is inserted through the skin (percutaneous) into an artery, usually in the groin, and then threaded to the blocked site in the coronary artery. When the balloon is expanded the plaque is pushed to the arterial walls, relieving the blockage and improving blood flow.

Soon after PCTA was introduced it was discovered that a large number of patients (30 to 50% within 6 months of an angioplasty) experience recurrent ischemia (restenosis) because the treated arteries re-narrow [25, 26]. Restenosis has several causes, including elastic recoil of the blood vessel, vascular remodeling and neointimal hyperplasia [25, 26]. In 1986 Sigwart et al. [27] introduced coronary stents, which provide primary stability to the dilated blood vessel (i.e., keep it open) (Fig. 1.2). Combined with appropriate therapies, stents significantly reduce the risks of restenosis and thrombosis [16, 28-30]. The placement of coronary stents (or "bare metal stents") became the standard treatment for obstructed coronary arteries [31].

However, implanted stents were frequently accompanied by in-stent neointimal hyperplasia, i.e. the excessive growth of scar tissue inside the stent [31, 32]. Considerable efforts have been undertaken to develop stents with active surfaces, to reduce such side-effects. Deployment of drug-eluting stents which deliver an anti-proliferative drug (e.g., sirolimus or paclitaxel) reduced the restenosis tendency, but have raised concerns about long-term impairment of endothelial response, late stent thrombosis and hypersensitivity reactions [28, 31]. In addition, the choice of drug and its optimal dosage is challenging, and once the

1 Introduction

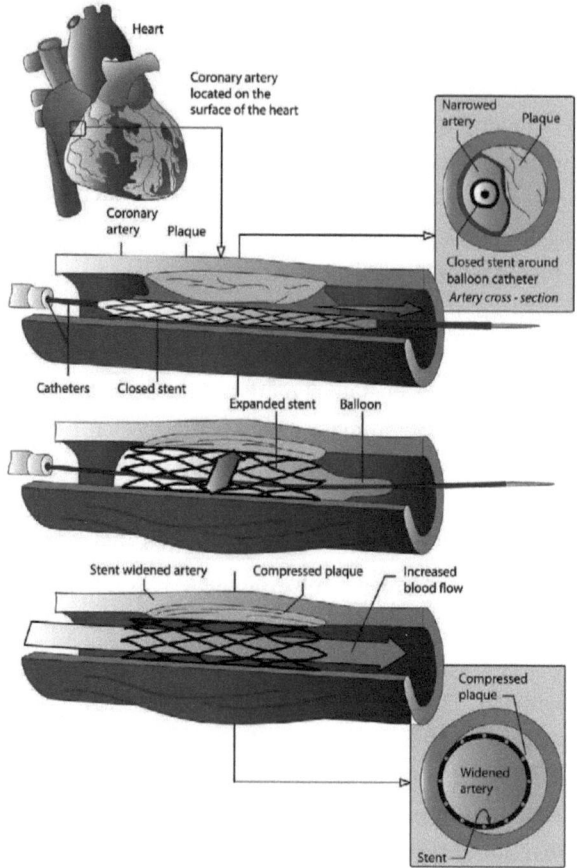

Figure 1.2: Schematic illustration showing PCTA with simultaneous stent placement. The stent is initially mounted on the balloon catheter and then expanded inside the artery (c.f. also [24]).

drug is completely released the long-term issues observed for bare metal stents also apply [31, 33]. Compared to the blood vessel a metallic stent is a rigid object, and relative motion between stent and vessel cause additional injuries. Finally, the permanent presence of a metallic object in the artery negatively influences diagnostic techniques such as magnetic resonance tomography [16].

1.2 Potential applications for degradable metals

Why, however, should a device implanted to prevent acute or subacute recoil during the initial period after PCTA, and to deliver anti-proliferative drugs over a few weeks, remain permanently in the body at all [16, 28, 34]? Temporary support of the narrowed blood vessel would circumvent the long-term risks and side effects mentioned above, and enable positive remodeling of the arterial tissue. Degradable implants are also especially interesting in pediatric cases (e.g., for children with congenital heart defects), where blood vessels are still growing and a fixed obstruction like a stent is a severe limitation [35].

1.2.2 Elastic stable intramedullary nailing

ESIN is a surgical procedure which was especially developed to treat pediatric bone fractures [12, 36]. It is indicated in the treatment of diaphyseal and certain metaphyseal/-epiphyseal fractures of long bones in children and young adults [37-39]. The above-average number of complaints associated with treatment of fractures in growing children points to the need for an alternative to the conservative approaches of locking plate osteosynthesis and external fixation [37, 38].

In the ESIN technique elastic nails are inserted into the medullary canal using a metaphyseal approach, in such a way as to stabilize the fractures [9, 39]. After skin incision two symmetrical drilling holes are usually made, and two nails are introduced into the medullary canal. After repositioning of the bone fragments, the nails are anchored at one end in the rather dense metaphyseal area of the bone, and at the other at the entry point [36, 38, 39]. The nails are usually pre-bent for a 3-point fixation of the bone, and by using two nails a balanced construct is achieved that maintains the correct alignment of the bone [39]. The fixation of a forearm (fractured radius and ulna) using ESIN is shown in Fig. 1.3.

The success of ESIN is in part based on the fact that it respects the particular biological characteristics of growing bone and the nature of children's fractures [39]. The periosteum of children is more biologically active, and cutting or stripping it has a deleterious effect on healing in terms of speed, callus formation and bone length [39]. ESIN minimizes injury to the periosteum, and the elasticity

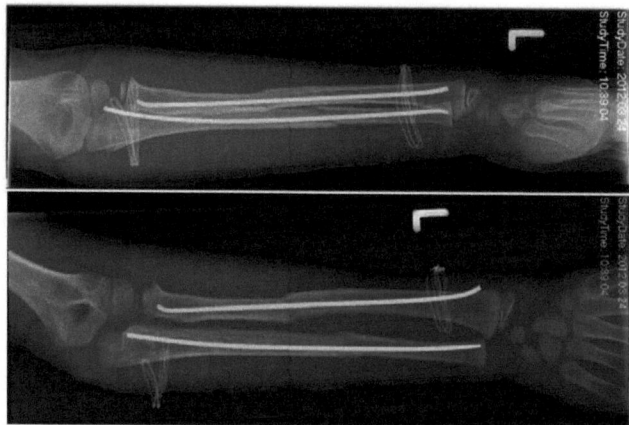

Figure 1.3: X-ray images of the ESIN fixation of a fractured left forearm. Images courtesy of T. Kraus and S. Fischerauer.

of the construct allows micro motion, aiding fast recovery [39]. The materials used in ESIN are either Ti alloys or stainless steel; following complete fracture healing, the implants have to be removed [36, 39]. A biodegradable material that provides sufficiently high strength and degrades in the body after fracture healing would circumvent the need for a second surgical intervention [12].

1.2.3 Further applications

Biodegradable metals are of interest not only for use in stents and ESIN but also for further applications in the contexts of (e.g.) osteosynthesis systems (plates, screws, nails, etc.) [11, 40, 41], wound closure [42, 43], treatment of tracheomalacia (a condition characterized by flaccidity of the tracheal support cartilage which may lead to tracheal collapse) [44], and intracranial aneurism (localized, blood-filled balloon-like bulge in the wall of an artery at the base of the brain) [45, 46].

In osteosynthesis, biodegradable metals are of particular interest in the areas of cranio-maxillofacial (CMF) [47], hand or podiatric surgery because (i) the implant sizes are comparatively small and hence the amount of material that has to degrade is limited, and (ii) second interventions are particularly cost-intensive, technically challenging or associated with aesthetic complications.

1.3 Iron as biodegradable material

Both polymeric and metallic materials possess the property profile required for the above-mentioned applications. Metals, however, are superior to polymers in mechanical performance [7, 8]. Among metals, Mg and Fe have both been considered for degradable implants because they are essential trace elements in the human body [6, 13, 14, 35, 41, 42, 48-75].

Various in vitro and in vivo studies have demonstrated Fe's potential for use as degradable implant material [6, 14, 35, 48, 53, 54, 70, 73-75]. In an initial in vivo study by Peuster et al. [14], stents produced from pure iron were implanted in the descending aorta of New Zealand white rabbits. The main findings were that no pronounced neointimal proliferation and no significant inflammatory response in the stented vessel occurred during the 18-month follow-up. A subsequent study over 12 months reported on the biocompatibility of Fe stents implanted in the descending aorta of minipigs [35]. It was concluded that Fe is a suitable material for degradable stents and does not cause local or systemic toxicity. However, it was also concluded that its overall degradation rate is too low, and materials with higher degradation rates were anticipated. A further investigation over 28 days by Waksman et al. [70] confirmed the results obtained in [35]. A study by Mueller et al. [76] also found that ferrous ions reduce the proliferation of smooth muscle cells and may therefore help to combat in-stent restenosis.

To increase the degradation rate of pure iron, Hermawan at al. [49, 77] developed an iron-manganese (Fe–Mn) alloy containing 35 wt.% Mn (Fe-35Mn) which degrades more quickly than pure iron [78]. Nevertheless, the degradation rate of Fe-35Mn is still at least one order of magnitude lower than that of magnesium alloys which have already been deployed in temporary stent applications [16, 28], and it is considered too slow for many of these. The following presents a

1 Introduction

design strategy [79] based on microstructural, electrochemical and toxicological considerations which addresses the limited mechanical performance and slow degradation behavior in temporary medical implants of the available Fe-based materials.

1.4 Design strategy for Fe-based degradable implant materials

The requirements of the applications mentioned above vary greatly. Stents, for example, require high-ductility materials with appropriate strength, while ESIN needs high-strength materials (especially high yield strength) with only moderate ductility. Only a versatile and effective design strategy can yield successful materials for such applications. Outstanding material performance also makes possible smaller dimensions, i.e. filigree implants which in turn reduce the amount of material degrading in the body.

The following sections present the considerations on which the design strategy relies and their consequences for material performance. Although divided into sections, the requirements formulated are linked, and the design strategy aims at optimal overall material performance. The intention of the following sections is to provide background information and comments to support the design strategy presentation in Chapter 2.

1.4.1 Electrochemical considerations

The approach to achieving increased degradation rates takes into account two criteria which influence the corrosion susceptibility of the metal:

1. The addition of less noble alloying elements within the solubility limit in Fe to make the Fe matrix more susceptible to corrosion.

2. The addition of noble alloying elements to generate small and finely dispersed precipitates that act as cathodic sites towards the Fe matrix, inducing microgalvanic corrosion.

1.4 Design strategy for Fe-based degradable implant materials

To satisfy criterion (1), the electromotive force (emf) series listed in Table 1.3 must be taken into account. This is an orderly arrangement of the standard potentials for all metals; the more negative values correspond to more reactive metals [80]. The position of a given element in the emf series is determined by the equilibrium potential of the metal in contact with an aqueous solution with its ions at a concentration equal to unit activity. Although this situation rarely occurs in practice, it is a useful aid in establishing a ranking among the different elements.

Table 1.3: Electromotive force series for selected elements [80].

Electrode reaction	Standard potential [V] at 25 °C
$Pt^{2+} + 2e^- = Pt$	≈ 1.2
$Pd^{2+} + 2e^- = Pd$	0.987
$Ag^+ + e^- = Ag$	0.800
$Cu^+ + e^- = Cu$	0.521
$Cu^{2+} + 2e^- = Cu$	0.342
$2H^+ + 2e^- = H_2$	0.000
$Sn^{2+} + 2e^- = Sn$	−0.136
$Ni^{2+} + 2e^- = Ni$	−0.250
$Co^{2+} + 2e^- = Co$	−0.277
$Fe^{2+} + 2e^- = Fe$	−0.440
$Cr^{3+} + 3e^- = Cr$	−0.74
$Zn^{2+} + 2e^- = Zn$	−0.763
$Cr^{2+} + 2e^- = Cr$	−0.91
$Mn^{2+} + 2e^- = Mn$	−1.18
$Ti^{2+} + 2e^- = Ti$	−1.63
$Al^{3+} + 2e^- = Al$	−1.66
$Mg^{2+} + 2e^- = Mg$	−2.37
$Li^+ + e^- = Li$	−3.05

1 Introduction

Several metals listed in the emf series are less noble than Fe. Most of these were not considered further, however, as either their solubility in Fe (e.g. Li, Mg, Ti) or their biocompatibility (e.g. Al, Cr) is limited. Mn was finally chosen to meet criterion (1) because it possesses a distinctly lower reduction potential ($E_{Mn} = -1.18$ V) than Fe ($E_{Fe} = -0.44$ V) and shows high solubility in Fe. Because Fe and Mn form a solid solution, the standard potential of the Fe–Mn alloy is expected to decrease with increasing Mn content [81]. The same approach was previously reported by Hermawan et al. in [49, 77] and was based on metallurgical and toxicological considerations. It corresponds to criterion (1) by considering Mn a suitable alloying element, and is clearly a step in the desired direction.

The potential of criterion (2) has already been illustrated in systems such as Al–Cu, where precipitates (Al_2Cu) are formed that are nobler than the Al matrix and thus reduce the alloy's corrosion resistance [82]. The efficiency of such an approach can be enhanced by reducing the size of the precipitates and distributing them homogeneously in the matrix. Here the degradation rate is expected to increase, while the material maintains homogeneous "macroscopic" degradation behavior.

The elements available for pursuing criterion (2) are all those in the emf series which are nobler than Fe (c.f. Table 1.3). The requirement that small and homogeneously distributed precipitates be formed imposes additional restrictions on the choice of element, as explained in the following section.

1.4.2 Formation of precipitates

In accordance with [79], the following requirements must be taken into account in selecting an appropriate alloying element:

1. It should have a limited and temperature-dependent solubility in Fe (or the Fe–Mn-matrix, respectively) to enable it to form precipitates.
2. To ensure that they are cathodic compared to the matrix, the precipitates should contain a large amount of the noble alloying element.

1.4 Design strategy for Fe-based degradable implant materials

3. Suitable process parameters are required to ensure homogeneous precipitate distribution and restricted precipitate size.

4. The influence of potential alloying elements on aspects of biocompatibility must be considered.

Based on a systematic evaluation of the ability of possible alloying elements to meet the above-mentioned requirements, the final choice was Pd. The considerations leading to this selection are described in detail in Chapter 2. From a metallurgy point of view this decision was fortunate. The precipitation tendency of Pd in the Fe–Mn-matrix is high and, given the appropriate microstructural characteristics of the matrix, precipitation can be readily triggered: Moszner et al. [83] demonstrated the rapid formation of Pd-rich precipitates in the martensitic Fe-10Mn-1Pd alloy. They showed that precipitates tend to form along dislocation lines via pipe diffusion and thereby demonstrate kinship with maraging steels. Following the same line of thought, the microstructure and mechanical performance of austenitic Fe–Mn–C–Pd alloys were optimized according to the design strategy (c.f. Chapter 3).

However, the choice of Pd also posed challenges. From the binary phase diagrams of Fe–Mn, Fe–Pd and Mn–Pd (c.f. Fig. 2.1), the solution heat treatment at a temperature of 1100 °C, as initially employed, was expected to yield a homogeneous single phase microstructure. The (surprising) presence of Pd-rich precipitates even in the solution-heat-treated state indicated their high thermal stability. In fact, further research ([83] and results presented in Chapter 4) showed that the precipitates' phase is probably MnPd with a high melting point of 1515 °C [84]. Consequently, they are expected to have a high enthalpy of formation and an extended stability range in the ternary system Fe–Mn–Pd, also towards very low Pd concentrations. Subsequently, the solution heat treatment temperature was increased to 1250 °C to achieve complete dissolution of the precipitates and a single phase microstructure [83].

1 Introduction

1.4.3 In vivo corrosion

As described in Section 1.1 the term (bio-) degradation is normally used in this thesis except where emphasis is laid on the electrochemical (corrosion) processes that take place during degradation – as here. In this context the in vivo situation represents a corrosive environment that is worth examining in detail. The first step is to consider the thermodynamics of the metal and the environment. M. Pourbaix [85, 86] developed a compact representation of thermodynamic data in the form of potential–pH diagrams that relate to the electrochemical and corrosion behavior of metals in water [80]. In their construction the various possible corrosion and product formation reactions are taken into account. Pourbaix diagrams show, at a glance the potential–pH domains in which the various contributing species (metal, ions etc.) are thermodynamically stable. The corresponding diagrams for Fe and Mn are shown in Fig. 1.4. They were calculated using the OLIAnalyzer software package (version 3.2.11, OLI Systems Inc.) for Fe and Mn in contact with simulated body fluid at 37 °C (c.f. Chapter 4 for the composition).

The two diagonal lines (labeled a and b) in the Pourbaix diagram delimit the thermodynamically stable region of water [80]. Above line b, oxygen evolves according to the reaction

$$H_2O \rightarrow \frac{1}{2}O_2 + 2H^+ + 2e^- \qquad (1.1)$$

Below line a, hydrogen gas evolves from the surface of an immersed electrode according to

$$2H^+ + 2e^- \rightarrow H_2 \qquad (1.2)$$

Between the two lines, water is stable and aqueous corrosion occurs in the region between these two lines [1]. For most metals, three different regions are present in a Pourbaix diagram: corrosion, passivity, and immunity. In the latter, the metal itself is stable and corrosion is energetically unfavorable. Passivity indicates that anodic metal dissolution takes place, but that there is a strong tendency to form corrosion products (usually termed passive layer) that cover

1.4 Design strategy for Fe-based degradable implant materials

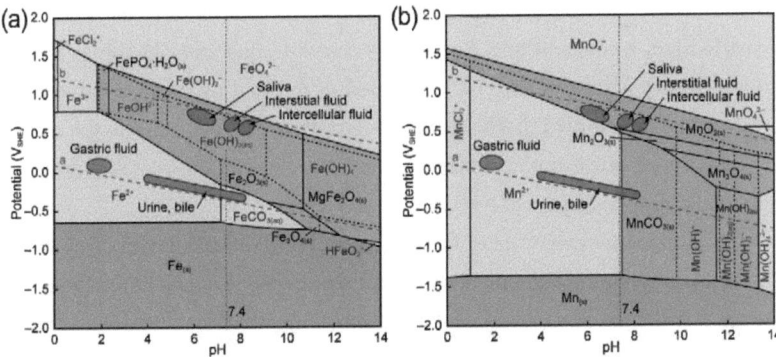

Figure 1.4: Calculated Pourbaix diagrams of (a) Fe and (b) Mn in simulated body fluid. All lines indicate thermodynamic equilibriums between two species. The solid lines delineate the regions of existence for solid or aqueous species, whereas dashed lines indicate boundaries of aqueous species that coexist with at least one solid phase. They are found only within the stability ranges of solids. The gray area shows the immunity of the metal, whereas the green region represents the passivity range. Blue areas stand for the regions in which only corrosion (no stable solid phases) occurs. The diagonal lines a and b represent the oxygen and hydrogen evolution reactions, respectively.

the surface and hence reduce the corrosion rate. Given that the passive layer is dense and stable, the corrosion process may even be completely inhibited, as e.g. for Al and its alloys. Upon corrosion (oxidation) of a metal, electrons are generated (c.f. the reactions listed in Table 1.3) which have to be taken up by a corresponding reduction reaction. The two reactions formulated above are both possible counter-reactions which balance the metal corrosion. Note that the reactions then proceed in the opposite (i.e., reductive) direction, as indicated above. Reaction (1.1) is usually written in the form

$$H_2O + \frac{1}{2}O_2 + 2e^- \rightarrow 2HO^- \qquad (1.3)$$

This representation corresponds more to the situation in neutral and slightly basic pH-values.

In the human body, saliva, intracellular fluid, plasma and interstitial fluids contain a high oxygen concentration [1]. The corresponding regions in the Pour-

1 Introduction

baix are therefore near the oxygen evolution line (b). In other areas, such as the gastric fluid, urine, and bile, the oxygen concentration and pH-values are different, as also shown in Fig. 1.4. In injured or infected tissue the pH-value can also change markedly [1]. The significance of Pourbaix diagrams in the context of metals in medical applications in general is to identify the overall behavior of the material with respect to its corrosion tendency.

Let us consider cases where Fe is in contact with blood plasma. Depending on the oxygen concentration, an equilibrium potential between Fe oxidation and oxygen reduction is achieved which lies above the metal stability ($Fe_{(s)}$) and below the oxygen reduction line (b); see Fig. 1.4a. The Pourbaix diagram shows that under these conditions it is predicted that Fe will be passive (formation of either Fe_2O_3, or $FeCO_3$). For Mn (Fig. 1.4b), the passive region (formation of $MnCO_3$) is restricted to slightly higher pH values and therefore Mn is expected to corrode actively.

It is known from in vivo studies [14, 35, 70], however, that the Fe stents implanted actually did corrode to a measureable extent. This observation indicates that even though Pourbaix diagrams are a good starting point for investigations on corrosion and highly valuable for equilibrium considerations, they also have limitations. The dynamic conditions in the human body, the influence of additional ions (e.g. Cl^- ions, which promote localized attacks), and further elements such as proteins and cells cannot be accurately represented [1]. The corrosion mechanism of Fe–Mn alloys in a physiological environment is discussed further in Chapter 4.

1.4.4 Biocompatibility considerations

The selected alloying elements not only influence the microstructure and mechanical properties, but may also significantly alter the biological performance of the material. In terms of biological performance the goal is to achieve an appropriate host response to the implantation of a degradable material. In the end it is the biological response which determines whether an implant material will be successful. However, this aspect is probably the most difficult to influence by

1.4 Design strategy for Fe-based degradable implant materials

means of material design. In the case of degradable materials in particular, the choice of alloying elements is important [60, 87, 88]. Both the base metal (Fe) and the additional alloying elements were chosen with respect to their biocompatibility.

Iron is essential for fundamental metabolic processes in cells and organisms [89]. The human body comprises approximately 40–50 mg·kg^{-1} body weight, most of it bound to hemoglobin (Table 1.4) [90]. Extracellular Fe is bound to transferrin, which both keeps Fe soluble and nontoxic [89, 91, 92] and makes plasma an extremely dynamic medium for Fe transport [90]. It is estimated that the transferrin molecule undergoes more than one hundred cycles of iron binding, transport and release before it is finally removed from circulation [90]. The Fe metabolism in humans is tightly regulated, and because there is no efficient mechanism for Fe excretion the daily Fe uptake usually only compensates for the corresponding losses. In the context of degrading implant materials, this means that if the degradation rate and hence the Fe release rate are not too high, the body is expected to tolerate the additional Fe. The maximum daily exposure limit for Fe lies in the range of milligrams per day (3 mg [88] to 40 mg [60]), depending on the assessment.

Table 1.4: Overview of Fe distribution in the human body [90].

Location	Amount [mg·kg^{-1} body weight]
Red blood cells (bound to hemoglobin)	30
Muscles (oxygen storage protein myoglobin)	4
Various tissues	2
Liver, spleen, bone marrow and muscle (as ferritin and haemosiderin)	10 – 12
Plasma (bound to transferrin)	3

Previous cell studies into pure Fe [48, 54, 73, 74], Fe–Mn [52, 63] and Fe–Mn–Si alloys [63] and the influence of common alloying elements in steels [62] showed that pure Fe possesses acceptable to good cytocompatibility [48, 54, 74]. How-

1 Introduction

ever, the alloys investigated also contain the alloying elements Mn, C and Pd. Even though Mn is a trace element which is essential for development and several body functions (e.g., bone formation and blood sugar regulation), and forms part of several enzymes, the body's tolerance limit for Mn is significantly lower than for Fe [88, 93]. The biocompatibility of Fe, Mn and Pd is discussed in detail in Chapter 6.

1.4.5 Aim of the project and outline of the thesis

The aim of this thesis project was the development and characterization of biodegradable Fe-based alloys for temporary medical implants. It was the intention to develop a novel alloy system characterized by both a faster degradation rate than that of currently proposed Fe-based alloys, and a high level of flexibility which would facilitate adjustments in mechanical performance. The latter achievement enabled the alloys in question to be deployed in the various applications described above.

The alloys developed in the study were characterized with respect to their microstructure, their degradation properties and their biological performance. Although the design strategy emphasizes the increase in the degradation rate, the alloys' microstructural characteristics were considered equally important. This was because a successful material has to fulfill the entire property profile of a particular application; in addition, it corresponds to the material scientist's approach, which is to control overall material performance via the design of optimal microstructure. This requires an understanding of the metal-physical processes that can be deployed to optimize the material.

These thoughts have influenced the structure of the thesis. In addition, the thesis is a cumulative work, and consists of five publications, which comprise Chapters 2–6. Chapter 2 presents the design strategy and preliminary results on biodegradable Fe-based alloys. Findings regarding the microstructure, mechanical properties and degradation performance of Fe-10Mn-1Pd are shown as an example. In Chapter 3 the design strategy is then extended to cover austenitic high-Mn steels, resulting in development of an alloy with the particular composition Fe-21Mn-0.7C-1Pd. This alloy is characterized by both high ductility and high strength values, and by means of an appropriate heat treatment its mechanical performance can be adjusted over a wide range. Chapter 4, following, is dedicated to assessing the degradation properties of the metals. Here a testing setup for immersion tests in simulated body fluid is described and the influence of the testing environment on the outcome of degradation studies is discussed. In Chapter 5 the degradation properties of the high-Mn alloys are determined using immersion tests and electrochemical techniques. Chapter 6 describes the biological response, investigated via in vitro cell studies. Finally, Chapter 7 summarizes the principal findings of the thesis and presents a brief outlook.

References

[1] Wong JY, Bronzino JD. Biomaterials. Boca Raton, FL: CRC Press; 2007.

[2] Bruck SD. Properties of biomaterials in the physiological environment. Boca Raton, FL: CRC Press; 1980.

[3] Wintermantel E, Ha S-W. Medizintechnik mit Biokompatiblen Werkstoffen. Third ed. Berlin: Springer; 2002.

[4] Black J. Biological Performance of Materials: Fundamentals of Biocompatibility. Boca Raton, FL: CRC Press; 2006.

[5] Williams DF. The Williams dictionary of Biomaterials. Liverpool: Liverpool University Press; 1999.

[6] Moravej M, Mantovani D. Biodegradable Metals for Cardiovascular Stent Application: Interests and New Opportunities. Int J Mol Sci 2011;12:4250-70.

[7] Mani G, Feldman MD, Patel D, Agrawal CM. Coronary stents: A materials perspective. Biomaterials 2007;28:1689-710.

[8] O'Brien B, Carroll W. The evolution of cardiovascular stent materials and surfaces in response to clinical drivers: A review. Acta Biomater 2009;5:945-58.

[9] Lascombes P, Haumont T, Journeau P. Use and Abuse of Flexible Intramedullary Nailing in Children and Adolescents. J Pediatr Orthop 2006;26:827-34.

[10] Lieber J, Härter B, Schmid E, Kirschner HJ, Schmittenbecher PP. Elastic Stable Intramedullary Nailing (ESIN) of Pediatric Metacarpal Fractures: Experiences with 66 Cases. Eur J Pediatr Surg 2012.

1 Introduction

[11] Staiger MP, Pietak AM, Huadmai J, Dias G. Magnesium and its alloys as orthopedic biomaterials: A review. Biomaterials 2006;27:1728-34.

[12] Celarek A, Kraus T, Tschegg EK, Fischerauer SF, Stanzl-Tschegg S, Uggowitzer PJ, et al. PHB, crystalline and amorphous magnesium alloys: Promising candidates for bioresorbable osteosynthesis implants? Mater Sci Eng C 2012;32:1503-10.

[13] Hänzi AC, Gerber I, Schinhammer M, Löffler JF, Uggowitzer PJ. On the in vitro and in vivo degradation performance and biological response of new biodegradable Mg–Y–Zn alloys. Acta Biomater 2010;6:1824-33.

[14] Peuster M, Wohlsein P, Brugmann M, Ehlerding M, Seidler K, Fink C, et al. A novel approach to temporary stenting: degradable cardiovascular stents produced from corrodible metal - results 6-18 months after implantation into New Zealand white rabbits. Heart 2001;86:563-9.

[15] Heublein B, Rohde R, Kaese V, Niemeyer M, Hartung W, Haverich A. Biocorrosion of magnesium alloys: a new principle in cardiovascular implant technology? Heart 2003;89:651-6.

[16] Di Mario C, Griffiths H, Goktekin O, Peeters N, Verbist J, Bosiers M, et al. Drug-eluting bioabsorbable magnesium stent. J Interv Cardiol 2004;17:391-5.

[17] Witte F, Hort N, Vogt C, Cohen S, Kainer KU, Willumeit R, et al. Degradable biomaterials based on magnesium corrosion. Curr Opin Solid State Mater Sci 2009;12:63-72.

[18] Murray CJL, Lopez AD. Mortality by cause for eight regions of the world: Global Burden of Disease Study. Lancet 1997;349:1269-76.

[19] Murray CJL, Lopez AD. Alternative projections of mortality and disability by cause 1990-2020: Global Burden of Disease Study. Lancet 1997;349:1498-504.

[20] Lopez AD, Mathers CD, Ezzati M, Jamison DT, Murray CJL. Global and regional burden of disease and risk factors, 2001: systematic analysis of population health data. Lancet 2006;367:1747-57.

[21] Ross R. The pathogenesis of atherosclerosis: a perspective for the 1990s. Nature 1993;362:801-9.

[22] Williams KJ, Tabas I. The Response-to-Retention Hypothesis of Early Atherogenesis. Arterioscler Thromb Vasc Biol 1995;15:551-61.

[23] Perrins C-J, Bobryshev Y. Current advances in understanding of immunopathology of atherosclerosis. Virchows Arch 2011;458:117-23.

[24] http://www.nhlbi.nih.gov/health/health-topics/topics/angioplasty/. What is

Coronary Angioplasty? Accessed 2012-06-20.

[25] Rajagopal V, Rockson SG. Coronary restenosis: a review of mechanisms and management. Am J Med 2003;115:547-53.

[26] Fischman DL, Leon MB, Baim DS, Schatz RA, Savage MP, Penn I, et al. A Randomized Comparison of Coronary-Stent Placement and Balloon Angioplasty in the Treatment of Coronary Artery Disease. New Engl J Med 1994;331:496-501.

[27] Sigwart U, Puel J, Mirkovitch V, Joffre F, Kappenberger L. Intravascular Stents to Prevent Occlusion and Re-Stenosis after Transluminal Angioplasty. New Engl J Med 1987;316:701-6.

[28] Erbel R, Di Mario C, Bartunek J, Bonnier J, de Bruyne B, Eberli FR, et al. Temporary scaffolding of coronary arteries with bioabsorbable magnesium stents: a prospective, non-randomised multicentre trial. Lancet 2007;369:1869-75.

[29] Serruys PW, de Jaegere P, Kiemeneij F, Macaya C, Rutsch W, Heyndrickx G, et al. A Comparison of Balloon-Expandable-Stent Implantation with Balloon Angioplasty in Patients with Coronary Artery Disease. New Engl J Med 1994;331:489-95.

[30] Mitra AK, Agrawal DK. In stent restenosis: bane of the stent era. J Clin Path 2006;59:232-9.

[31] Serruys PW, Kutryk MJB, Ong ATL. Drug therapy - Coronary-artery stents. New Engl J Med 2006;354:483-95.

[32] Hoffmann R, Mintz GS. Coronary in-stent restenosis - predictors, treatment and prevention. Eur Heart J 2000;21:1739-49.

[33] Bennett MR. In-stent Stenosis: Pathology and Implications for the Development of Drug Eluting Stents. Heart 2003;89:218-24.

[34] Colombo A, Karvouni E. Biodegradable Stents : "Fulfilling the Mission and Stepping Away". Circulation 2000;102:371-3.

[35] Peuster M, Hesse C, Schloo T, Fink C, Beerbaum P, von Schnakenburg C. Long-term biocompatibility of a corrodible peripheral iron stent in the porcine descending aorta. Biomaterials 2006;27:4955-62.

[36] Weinberg A-M, Castellani C, Amerstorfer F. Elastisch-stabile intramedulläre Marknagelung (ESIN) von Unterarmfrakturen. Oper Orthop Traumatol 2008;20:285-96.

[37] Kraus R, Wessel L. The Treatment of Upper Limb Fractures in Children and Adolescents. Dtsch Arztebl International 2010;107:903-10.

[38] Dietz HG, Schlickewei W. Femurschaftfrakturen im Kindesalter. Unfallchirurg 2011;114:382-7.

1 Introduction

[39] Hunter JB. The principles of elastic stable intramedullary nailing in children. Inj 2005;36:S20-S4.

[40] Witte F, Kaese V, Haferkamp H, Switzer E, Meyer-Lindenberg A, Wirth CJ, et al. In vivo corrosion of four magnesium alloys and the associated bone response. Biomaterials 2005;26:3557-63.

[41] Kraus T, Fischerauer SF, Hänzi AC, Uggowitzer PJ, Löffler JF, Weinberg AM. Magnesium alloys for temporary implants in osteosynthesis: In vivo studies of their degradation and interaction with bone. Acta Biomater 2012;8:1230-8.

[42] Hänzi AC, Metlar A, Schinhammer M, Aguib H, Lüth TC, Löffler JF, et al. Biodegradable wound-closing devices for gastrointestinal interventions: Degradation performance of the magnesium tip. Mater Sci Eng C 2011;31:1098-103.

[43] Shishatskaya EI, Volova TG, Puzyr AP, Mogilnaya OA, Efremov SN. Tissue response to the implantation of biodegradable polyhydroxyalkanoate sutures. J Mater Sci: Mater Med 2004;15:719-28.

[44] Austin J, Ali T. Tracheomalacia and bronchomalacia in children: pathophysiology, assessment, treatment and anaesthesia management. Pediatr Anesth 2003;13:3-11.

[45] Hirabayashi M, Ohta M, Rüfenacht DA, Chopard B. A lattice Boltzmann study of blood flow in stented aneurism. Future Gener Comput Syst 2004;20:925-34.

[46] Alpagut U, Ugurlucan M, Kafali E, Surmen B, Sayin OA, Guven K, et al. Endoluminal stenting of mycotic saccular aneurysm at the aortic arch. Tex Heart Inst J 2006;33:371-5.

[47] Yun Y, Dong Z, Yang D, Schulz MJ, Shanov VN, Yarmolenko S, et al. Biodegradable Mg corrosion and osteoblast cell culture studies. Mater Sci Eng C 2009;29:1814-21.

[48] Nie FL, Zheng YF, Wei SC, Hu C, Yang G. In vitro corrosion, cytotoxicity and hemocompatibility of bulk nanocrystalline pure iron. Biomed Mater 2010;5.

[49] Hermawan H, Alamdari H, Mantovani D, Dube D. Iron-manganese: new class of metallic degradable biomaterials prepared by powder metallurgy. Powder Metall 2008;51:38-45.

[50] Hermawan H, Dube D, Mantovani D. Degradable metallic biomaterials: Design and development of Fe–Mn alloys for stents. J Biomed Mater Res, Part A 2010;93A:1-11.

[51] Hermawan H, Dubé D, Mantovani D. Developments in metallic biodegradable stents. Acta Biomater 2010;6:1693-7.

[52] Hermawan H, Purnama A, Dube D, Couet J, Mantovani D. Fe–Mn alloys for metallic biodegradable stents: Degradation and cell viability studies. Acta Biomater 2010;6:1852-60.

[53] Moravej M, Prima F, Fiset M, Mantovani D. Electroformed iron as new biomaterial for degradable stents: Development process and structure-properties relationship. Acta Biomater 2010;6:1726-35.

[54] Moravej M, Purnama A, Fiset M, Couet J, Mantovani D. Electroformed pure iron as a new biomaterial for degradable stents: In vitro degradation and preliminary cell viability studies. Acta Biomater 2010;6:1843-51.

[55] Purnama A, Hermawan H, Couet J, Mantovani D. Assessing the biocompatibility of degradable metallic materials: State-of-the-art and focus on the potential of genetic regulation. Acta Biomater 2010;6:1800-7.

[56] Choudhary L, Singh Raman RK. Magnesium alloys as body implants: Fracture mechanism under dynamic and static loadings in a physiological environment. Acta Biomater 2012;8:916-23.

[57] Gunde P, Hänzi AC, Sologubenko AS, Uggowitzer PJ. High-strength magnesium alloys for degradable implant applications. Mater Sci Eng A 2010;528:1047-54.

[58] Hänzi AC, Gunde P, Schinhammer M, Uggowitzer PJ. On the biodegradation performance of an Mg–Y–RE alloy with various surface conditions in simulated body fluid. Acta Biomater 2009;5:162-71.

[59] Hänzi AC, Sologubenko AS, Uggowitzer PJ. Design Strategy for Microalloyed Ultra-Ductile Magnesium Alloys for Medical Applications. Light Met Technol 2009;618-619:75-82.

[60] Kirkland N, Staiger M, Nisbet D, Davies C, Birbilis N. Performance-driven design of Biocompatible Mg alloys. JOM 2011;63:28-34.

[61] Li Z, Gu X, Lou S, Zheng Y. The development of binary Mg–Ca alloys for use as biodegradable materials within bone. Biomaterials 2008;29:1329-44.

[62] Liu B, Zheng YF. Effects of alloying elements (Mn, Co, Al, W, Sn, B, C and S) on biodegradability and in vitro biocompatibility of pure iron. Acta Biomater 2011;7:1407-20.

[63] Liu B, Zheng YF, Ruan L. In vitro investigation of Fe30Mn6Si shape memory alloy as potential biodegradable metallic material. Mater Lett 2011;65:540-3.

[64] Liu C, Xin Y, Tian X, Chu PK. Degradation susceptibility of surgical magnesium alloy in artificial biological fluid containing albumin. J Mater Res 2007;22:1806-14.

1 Introduction

[65] Liu CL, Wang YJ, Zeng RC, Zhang XM, Huang WJ, Chu PK. In vitro corrosion degradation behaviour of Mg–Ca alloy in the presence of albumin. Corros Sci 2010;52:3341-7.

[66] Mueller W-D, Fernández Lorenzo de Mele M, Nascimento ML, Zeddies M. Degradation of magnesium and its alloys: Dependence on the composition of the synthetic biological media. J Biomed Mater Res, Part A 2009;90A:487-95.

[67] Peuster M, Beerbaum P, Bach F-W, Hauser H. Are resorbable implants about to become a reality? Cardiol Young 2006;16:107-16.

[68] Pierson D, Edick J, Tauscher A, Pokorney E, Bowen P, Gelbaugh J, et al. A simplified in vivo approach for evaluating the bioabsorbable behavior of candidate stent materials. J Biomed Mater Res Part B 2012;100B:58-67.

[69] Waksman R. Update on Bioabsorbable Stents: From Bench to Clinical. J Interv Cardiol 2006;19:414-21.

[70] Waksman R, Pakala R, Baffour R, Seabron R, Hellinga D, Tio FO. Short-term effects of biocorrodible iron stents in porcine coronary arteries. J Interv Cardiol 2008;21:15-20.

[71] Waksman R, Pakala R, Kuchulakanti PK, Baffour R, Hellinga D, Seabron R, et al. Safety and efficacy of bioabsorbable magnesium alloy stents in porcine coronary arteries. Cathet Cardiovasc Interv 2006;68:607-17.

[72] Peeters P, Bosiers M, Verbist J, Deloose K, Heublein B. Preliminary results after application of absorbable metal stents in patients with critical limb ischemia. J Endovasc Ther 2005;12:1-5.

[73] Zhu S, Huang N, Xu L, Zhang Y, Liu H, Sun H, et al. Biocompatibility of pure iron: In vitro assessment of degradation kinetics and cytotoxicity on endothelial cells. Mater Sci Eng C 2009;29:1589-92.

[74] Zhang EL, Chen HY, Shen F. Biocorrosion properties and blood and cell compatibility of pure iron as a biodegradable biomaterial. J Mater Sci: Mater Med 2010;21:2151-63.

[75] Wegener B, Sievers B, Utzschneider S, Müller P, Jansson V, Rößler S, et al. Microstructure, cytotoxicity and corrosion of powder-metallurgical iron alloys for biodegradable bone replacement materials. Mater Sci Eng B 2011;176:1789-96.

[76] Mueller PP, May T, Perz A, Hauser H, Peuster M. Control of smooth muscle cell proliferation by ferrous iron. Biomaterials 2006;27:2193-200.

References

[77] Hermawan H, Dubé D, Mantovani D. Development of Degradable Fe-35Mn Alloy for Biomedical Application. Adv Mater Res 2007;15-17:107-12 [THERMEC 2006 Supplement].

[78] Hermawan H, Moravej M, Dubé D, Fiset M, Mantovani D. Degradation Behaviour of Metallic Biomaterials for Degradable Stents. Adv Mater Res 2007;15-17:113-8 [THERMEC 2006 Supplement].

[79] Schinhammer M, Hänzi AC, Löffler JF, Uggowitzer PJ. Design strategy for biodegradable Fe-based alloys for medical applications. Acta Biomater 2010;6:1705-13.

[80] Revie WR, Uhlig HH. Corrosion and Corrosion Control. Fourth ed: John Wiley & Sons; 2008.

[81] Kawashima A, Asami K, Hashimoto K. Effect of Manganese on the Corrosion Behaviour of Chromium-Bearing Amorphous Metal-Metalloid Alloys. Sci Rep Res Inst Tohoku Univ Phys Chem Metall 1981;29:276-83.

[82] Vargel C, Jacques M, Schmidt MP. Corrosion of Aluminium: Elsevier; 2004.

[83] Moszner F, Sologubenko AS, Schinhammer M, Lerchbacher C, Hänzi AC, Leitner H, et al. Precipitation hardening of biodegradable Fe-Mn-Pd alloys. Acta Mater 2011;59:981-91.

[84] Predel B. Landolt-Börnstein. In: Madelung O, editor. Group IV - Physical Chemistry: Springer-Verlag; 1998.

[85] Pourbaix M. Atlas d'équilibres électrochimiques. Paris: Gauthier-Villars; 1963.

[86] Pourbaix M. Electrochemical corrosion of metallic biomaterials. Biomaterials 1984;5:122-34.

[87] Brar H, Keselowsky B, Sarntinoranont M, Manuel M. Design considerations for developing biodegradable and bioabsorbable magnesium implants. JOM 2011;63:100-4.

[88] Yuen CK, Ip WY. Theoretical risk assessment of magnesium alloys as degradable biomedical implants. Acta Biomater 2010;6:1808-12.

[89] Hentze MW, Muckenthaler MU, Galy B, Camaschella C. Two to Tango: Regulation of Mammalian Iron Metabolism. Cell 2010;142:24-38.

[90] Crichton R. Inorganic Biochemistry of Iron Metabolism: From Molecular Mechanisms to Clinical Consequences. Second ed. New York: John Wiley & Sons; 2001.

[91] Halliwell B, Gutteridge JMC. Biologically relevant metal ion-dependent hydroxyl radical generation An update. FEBS Lett 1992;307:108-12.

1 Introduction

[92] Papanikolaou G, Pantopoulos K. Iron metabolism and toxicity. Toxicol Appl Pharmacol 2005;202:199-211.

[93] Yamamoto A, Honma R, Sumita M. Cytotoxicity evaluation of 43 metal salts using murine fibroblasts and osteoblastic cells. J Biomed Mater Res 1998;39:331-40.

2 Design strategy

This chapter presents the design strategy for the development of biodegradable Fe-based alloys for temporary medical implants. This design strategy is based on the considerations elucidated in the introduction and aims to optimize overall alloy performance in terms of microstructural features, mechanical properties, and degradation.

Design strategy for biodegradable Fe-based alloys for medical applications[1]

The aim of this article is to describe a design strategy for the development of new biodegradable Fe-based alloys offering a performance considered appropriate for temporary implant applications including both, enhanced degradation rate compared to pure iron, and suitable strength and ductility. The design strategy is based on electrochemical, microstructural and toxicological considerations. The influence of alloying elements on the electrochemical modification of the Fe-matrix and the controlled formation of noble intermetallic phases is deployed. Such intermetallic phases are responsible for both, an increased degradation rate and enhanced strength. Manganese and palladium have shown to be suitable alloying additions for this design strategy; Mn lowers the standard electrode potential while Pd forms noble (Fe,Mn)Pd intermetallics that act as cathodic sites. We discuss the efficiency and the potential of the design approach, and evaluate the resulting characteristics of the new alloys using metal-physical experiments including electrochemical measurements, phase identification analysis and electron microscopy studies. The newly-developed Fe–Mn–Pd alloys reveal a degradation resistance, which is one order of magnitude lower than observed for pure iron. Additionally, the mechanical performance is shown to be adjustable not only by the choice of alloying elements but also by heat treatment procedures; high strength values > 1400 MPa at ductility levels > 10% can be achieved. Thus, the new alloys offer an attractive combination of electrochemical and mechanical characteristics considered suitable for biodegradable medical applications.

2.1 Introduction

Interest in biodegradable metallic materials for use as temporary implant material in vascular intervention and osteosynthesis has been continuously increas-

---

[1]M. Schinhammer, A.C. Hänzi, J.F. Löffler, P.J. Uggowitzer; Acta Biomaterialia 6 (2010) 1705-13

2.1 Introduction

ing over the past few years [1-7]. It has been found that in certain cases degradable implants may overcome the disadvantages of permanent devices, such as prolonged physical irritation and chronic inflammation. Magnesium and its alloys are considered to be promising candidates for such applications, and they have already been successfully tested in vivo and in clinical studies [8-11]. In physiological media they exhibit appropriate or rapid degradation rates accompanied by the evolution of considerable amounts of hydrogen. The latter, however, can be problematic for a series of biomedical applications, especially in osteosynthesis where implants are generally larger; and the mechanical properties of Mg-based alloys, such as strength and ductility (specifically the elongation to fracture), are not always satisfactory.

Iron is considered to be an alternative interesting candidate for use as biodegradable implant material [6, 7, 12-14]. Preliminary in vivo studies have already shown the potential of iron for degradable medical applications: stents made of pure iron implanted into the porcine aorta did not induce any local or systemic toxicity [12]. However, because of the very low degradation rate of pure iron in physiological media such implants are considered to reveal reactions similar to those found in permanent applications [12]. In order to increase the degradation rate of Fe-based materials Hermawan et al. [6, 7] developed an Fe–Mn alloy containing 35 wt.% Mn (Fe-35Mn) which exhibits an increased degradation rate with respect to pure iron [5]. Nevertheless, compared to magnesium alloys, the degradation rate of Fe-35Mn is still at least one order of magnitude lower and considered too slow for many temporary implant applications. In this study we focus on cardiovascular stent applications because demand for such devices is high, as cardiovascular diseases are the most common cause of morbidity and mortality in the Western world. From a medical point of view a degradation period over 12–24 months would be advisable for such stent applications [15]. Consequently, the ideal degradation rate lies between that of Mg-alloys and the Fe-35Mn alloy. From a mechanical point of view the properties of stainless steel of type 316L are at least aimed at. This material is often used for permanent implant applications, exhibiting a yield stress of about 230 MPa at high elongation to fracture values [7]. The mechanical properties of powder-metallurgically processed Fe-35Mn are comparable to 316L steel. Higher yield strength at comparable elongation to fracture, however, would even allow a reduction of implant dimension which would be beneficial for the patient because the amount of material released in the body is smaller while offering the same support to the tissue. Taking these considerations into account for use as degradable stent materials it would be desirable to find Fe-based alloys with adapted degradation performance towards increased corrosion susceptibility and with improved mechanical properties regarding strength and ductility as compared to currently available Fe-alloys.

In this study we present a design strategy deployed to develop new Fe-based alloys for biodegradable stent applications with tailorable microstructural, electrochemical and mechanical properties. We discuss the efficiency of the design approach and present preliminary results on the degradation behavior and the microstructural and mechanical properties of the Fe-based alloys developed.

2.2 Design strategy

Results from previous studies [5-7, 10, 12] not only reveal the potential of Fe-based alloys as biodegradable implant material but also emphasize the need to increase their corrosion rate to meet the requirements of degradable stent applications. Therefore, our design strategy for Fe-based alloys elucidated in the following aims at both an increase in the degradation rate and an improvement of mechanical properties by modifying the chemical composition and microstructural characteristics in relation to currently known Fe-based alloys.

The approach to achieving increased degradation rate takes into account two criteria which influence the corrosion susceptibility of the metal: (1) the addition of less noble alloying elements within the solubility limit in Fe to make the Fe-matrix more susceptible to corrosion; and (2) the addition of noble alloying elements to generate small and finely dispersed intermetallic phases (IMPs) that act as cathodic sites towards the Fe-matrix, inducing microgalvanic corrosion. The potential of criterion (2) has already been illustrated in systems such as Al–Cu,

2.2 Design strategy

where precipitates (Al$_2$Cu) are formed that are nobler than the Al-matrix and thus decrease the alloy's corrosion resistance [16]. The efficiency of such an approach can be enhanced by reducing the size of the IMPs and achieving a homogeneous distribution of the IMPs in the matrix. In this case the degradation rate is expected to increase tremendously while the material maintains homogeneous degradation.

The approach described by Hermawan et al. [6, 7], which is based on metallurgical and toxicological considerations, pursues criterion (1) considering Mn as a suitable alloying element, and is clearly a step in the desired direction. Mn lowers the standard electrode potential (E^0) of iron: the standard potential of the reaction Fe $\rightarrow$ Fe^{2+} + 2e$^-$ is E_1 = -0.440 V [17], whereas that of Mn $\rightarrow$ Mn^{2+} + 2e$^-$ is E_2 = -1.18 V [17]. Because Fe and Mn form a solid solution [18] (see also Fig. 2.1a) the standard potential of the Fe–Mn alloy is expected to decrease with increasing Mn content [19]; this is taken up here to meet criterion (1).

The choice of alloying elements available to take into account criterion (2) is quite wide. However, several aspects must be considered: (i) the alloying elements should have a limited and temperature-dependent solubility in Fe such that they are likely to form IMPs; (ii) these IMPs should contain high amounts of the noble element to ensure that they are cathodic compared to the Fe-matrix; (iii) there have to be process parameters suitable for restricting the size of the intermetallic particles and for generating a fine dispersion of the intermetallic particles in the matrix; and (iv) the influence of potential alloying elements on aspects of biocompatibility need to be considered.

A systematic evaluation of possible alloying elements to meet the aspects (i)–(iv) resulted in the selection of Pd, not least because the standard potential of the reaction Pd $\rightarrow$ Pd^{2+} + 2e$^-$ is quite high, E_3 = -0.987 V [17], making Pd-containing IMPs likely to be nobler than the Fe-matrix. Further reasons for selecting Pd are discussed in the following sections, and are based partly on the binary phase diagrams (PDs) of Fe–Mn, Fe–Pd and Mn–Pd because information on the ternary Fe–Mn–Pd PDs is not available. In addition, only small amounts of Pd are considered to meet aspect (iii). As to meeting aspect (iv), it is noted that Pd is frequently used in casting alloys for dental restorations in orthodontic

2 Design strategy

applications [20]. Even though only little information on the effects of subcutaneously implanted Pd is available, it is considered to be tolerated by the body [21]. Given the small weight of a coronary Fe-stent ($\approx$ 20 mg), its relatively long degradation period and a content of less than 1 at.% Pd in the alloys suggested here, the Pd content is not expected to be problematic.

The binary PDs of Fe–Pd and Mn–Pd indicate that potential binary IMPs consist of at least 50 at.% Pd. In Fig. 2.1, however, only the Fe- and Mn-rich parts of the binary PDs are displayed, as they are of interest in this study according to the above considerations [18]. The Fe–Pd PD in Fig. 2.1b shows that at temperatures above 815 °C Fe and Pd are completely miscible. At lower temperatures the IMP FePd forms. Fig. 2.1c shows the corresponding Mn-rich part of the Mn–Pd PD. Again, at temperatures above 800 °C Mn and Pd form a solid solution over a wide Pd range, whereas at lower temperatures only a small amount of Pd is soluble in Mn and the β-phase, an IMP with the chemical composition MnPd, forms. We assume a temperature-dependent solubility of Pd in Fe and Mn to be likely also in the ternary system. This will be utilized to form Pd-rich IMPs from a supersaturated solid solution (SSSS) quenched from elevated temperatures (estimated at approximately 1100 °C) via an aging procedure.

Noting the influence of the selected alloying elements Mn and Pd on the mechanical properties the following has to be taken into account: in the Fe–Mn system at Mn contents below approximately 30 wt.% a phase transformation from the austenitic to the ferritic structure occurs upon cooling (Fig. 2.1a). This transformation requires significant changes in the phase compositions. Upon rapid cooling from elevated temperatures (above 900 °C), diffusion is limited and martensitic phase transformations do occur [22]. Consequently the system Fe–Mn(–Pd) exhibits a pronounced transformation hysteresis. The "real" PD in Fig. 2.2 [18] shows the temperature ranges for the martensitic and the reverse transformations ($\gamma \to \alpha'$, $\alpha' \to \gamma$, $\gamma \to \epsilon$, $\epsilon \to \gamma$, $\epsilon \to \alpha'$) which are important for determining appropriate heat treatment parameters. The types and amounts of phases, in turn, have an impact on the mechanical performance of the alloys. The austenitic phase (γ) enhances ductility, while the martensitic phase (α') generates hardening of the microstructure.

2.2 Design strategy

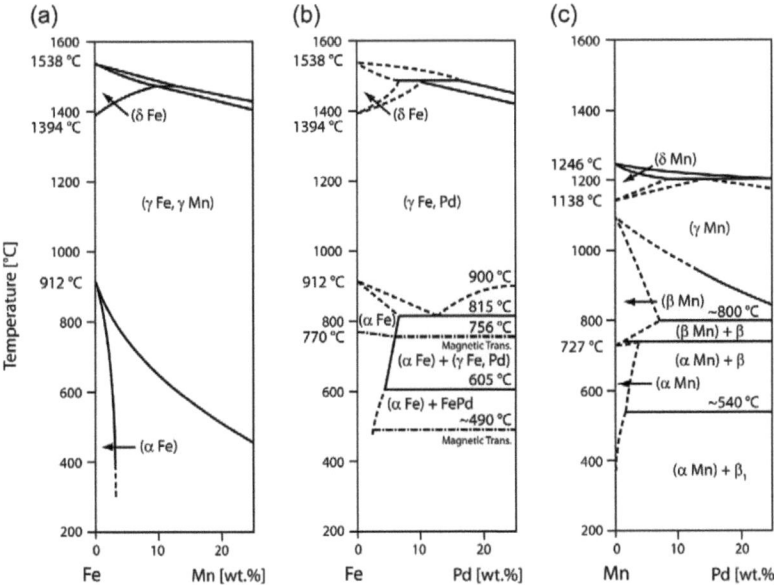

Figure 2.1: Sections of binary phase diagrams: (a) Fe-rich part of the Fe–Mn phase diagram; (b) Fe-rich part of the Fe–Pd phase diagram; (c) Mn-rich part of the Mn–Pd phase diagram.

The IMPs mentioned above can also be used to deploy a positive influence on the mechanical properties. Again, this will be illustrated using the system Al–Cu. Depending on the aging conditions, the size and the distance between the precipitates can be adjusted. This in turn may generate considerable impact on the strength of the alloy, taking into account strengthening according to the precipitation hardening mechanism [23].

According to the considerations illustrated above, an Fe-10Mn-1Pd alloy (in wt.%) was produced and the influence of the chemical composition on the electrochemical performance and on microstructural and mechanical properties was evaluated. It is noteworthy that in preliminary studies the Pd content was varied systematically with the finding that 1 wt.% Pd is sufficient to form the desired Pd-containing IMPs.

2 Design strategy

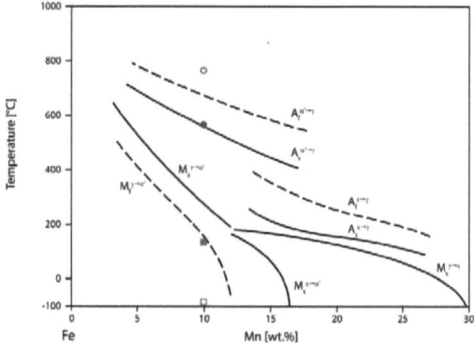

Figure 2.2: Martensitic transformation temperatures for the metastable α'- and ϵ-phases: starting (M_s) and finishing (M_f) temperatures on cooling, and starting (A_s) and finishing (A_f) temperatures on heating. The starting and finishing temperatures determined by dilatometry for the alloy studied here are also indicated.

2.3 Materials and methods

2.3.1 Sample preparation and heat treatments

The Fe-10Mn-1Pd (in wt.%) alloy was produced from the following raw materials: commercially available low-carbon steel (Alloy no. 1.0401) obtained from Metaltec AG (Switzerland), pure manganese (99.9%, Alfa Aesar, Germany) and palladium (99.95%, UBS, Switzerland). Beside this alloy, an alloy containing 10 wt.% Mn without Pd was produced. Pure iron (Armco) and low-carbon steel were selected as references and used in the as-received state. The compositions of the alloys measured by spark spectrometry are summarized in Table 2.1.

The elements were melted in a vacuum-induction furnace under 300 mbar argon atmosphere (99.998% purity) and cast into a copper mold. The cylindrical rods, with an initial diameter of 35 mm and a length of 100 mm, were forged at 1000 °C and subsequently swaged into rods of 12 mm in diameter. The alloys were then solution-heat-treated (sht) for 2 h at 1100 °C under argon atmosphere (99.998% purity) and subsequently water quenched. After the solution heat treatment, the samples were subjected to two different heat treatments, respectively, to study the influence of annealing parameters on the formation ten-

2.3 Materials and methods

Table 2.1: Composition of the alloys evaluated (in wt.%).

Designation	Fe	C	Mn	Pd	Si	P	S
Fe (Armco)	99.9	—	—	—	—	—	—
Fe (carbon steel)	Bal.	0.14	0.6	—	0.17	0.01	0.02
Fe-10Mn	Bal.	0.13	10.8	—	0.14	0.01	0.02
Fe-10Mn-1Pd	Bal.	0.12	10.2	0.92*	0.15	0.01	0.02

* Measured using ICP-OES analysis.

dencies of IMPs and the impact on microstructural, electrochemical and mechanical properties. In this study, two different heat treatments are shown exemplarily as part of an ongoing work on the physical characteristics of the precipitation behavior of this complex system.

Based on the information on the martensitic transformations given in Fig. 2.2, the following heat treatment parameters were chosen: (1) the alloy was aged for 1 h at 500 °C to achieve a microstructure consisting of tempered martensite, retained austenite and Pd-IMP precipitates (denoted as ht 1); (2) the alloy was briefly annealed in the γ-phase region (10 min at 700 °C) and subsequently aged for 10 h at 500 °C to achieve IMP formation in a state with a high amount of austenitic phase (denoted as ht 2). Both heat treatments were performed in air.

Cooling from the respective aging temperature was performed by water quenching. The heat treatments at 700 °C and 1100 °C were performed using an oven from Lenton Thermal Design Ltd (England), whereas the heat treatments at 500 °C were carried out in an air-circulating oven (Heraeus 170/2, Germany).

Samples for immersion testing were dressed to a diameter of 10 mm and cut into disks of 3 mm thickness. The entire disk surface was ground and polished (down to 3 μm diamond polish) and cleaned in ethanol in an ultrasonic bath prior to immersion.

2.3.2 Microstructure and characterization

The microstructures of all samples were investigated using optical microscopy (OM). For phase identification X-ray diffraction (XRD) measurements (PANalytical X'Pert PRO-MPD) were conducted using a Cu $K_{\alpha 1}$ (λ = 0.15406 nm) source operated at 37 kV and 45 mA. A Fischerscope mms (multi-measuring system, H. Fischer, Germany) equipped with a magneto-inductive probe was used to determine the ferrite content (in volume percent) of the samples.

Transmission electron microscopy (TEM) and energy dispersive X-ray analysis (EDX) were performed on a Philips CM30 device operated at 300 kV. Scanning electron microscopy (SEM) was done using a LEO-1530 device (Zeiss, Germany) equipped with a field emission gun operated at 20 kV. For the determination of the phase transformation temperatures dilatometry measurements were performed using a DIL 805 A dilatometer (Bähr-Thermoanalyse, Germany). A Brickers 220 (Gnehm, Switzerland) hardness measurement device was used to perform hardness measurements. All hardness values are given in Vickers units (HV10, indentation time 6 s).

2.3.3 Immersion testing and impedance spectroscopy

Degradation performance was evaluated by immersion testing in simulated body fluid (SBF). SBF is an aqueous solution that mimics the ionic composition and the pH value of human blood but does not contain larger particles such as proteins, lipids or blood cells. The composition of the SBF in this study was taken from [24], because it simulates the composition of human blood most accurately. For immersion testing, the samples were suspended in a beaker filled with SBF maintained at a constant temperature of 37 ± 1 °C in ambient air. The solution was stirred during the experiment to ensure constant solution conditions. The degradation rate was determined using the weight-loss method. Therefore the samples were weighed and their dimensions exactly determined prior to the experiment. After immersion the samples were cleaned in acetone in an ultrasonic bath and the remaining corrosion products were removed using a small brush. Finally, the samples were again rinsed with acetone before they were weighed the second time. Carbon steel and Fe-10Mn (ht 2) were selected as reference for the degradation performance. As comparison, the Fe-10Mn-1Pd alloy in ht 2 condition was chosen to represent the new Pd-containing steel in the aged condition.

Electrochemical impedance spectroscopy (EIS) was performed using an Autolab PGSTAT302 equipped with an additional FRA2 module (Eco Chemie B.V., The Netherlands). The samples were measured in a three-electrode setup: the sample acted as working electrode, a platinum sheet was used as counter electrode, and a saturated calomel electrode (SCE) was taken as reference electrode. The measurement was done in SBF over a frequency range of 10^5 to $5 \cdot 10^{-3}$ Hz, using the single AC mode with an amplitude of 10 mV.

The results of the EIS measurements are represented using Bode plots. They show the impedance and the phase angle as a function of the frequency of the applied alternating current. At the upper frequency limit the impedance value represents the resistance of the solution, whereas at the lower frequency limit the impedance values approach the sum of the polarization resistance and the solution resistance [25]. The polarization resistance is inversely proportional to the corrosion rate and therefore measures the corrosion susceptibility of the sample [25].

2.3.4 Mechanical testing

The mechanical properties were determined by standard tensile testing (Schenck-Trebel, Germany) of round specimens of 3 mm diameter and a gauge length of 15 mm. The tests were performed at an initial strain rate of 10^{-3} s^{-1}.

2.4 Results

2.4.1 Microstructure

Here the influence of the heat treatments on the relative amounts of the different phases according to XRD analysis is presented first, to aid understanding of the following microstructural observations using OM, SEM and TEM. In Fig. 2.3, in addition to the two phases already described (α' and γ), a third phase, the ϵ-phase (ϵ-martensite), is seen in the XRD-spectra. This phase has a hexagonal close-packed (hcp) structure and usually only occurs at Mn contents above 12 wt.% [18]. However, ϵ-phase is detected even in the sht condition. The spectrum of the sample in the ht 1 condition differs only slightly from that in the sht state: the relative intensity of the γ-phase peaks is slightly lower. In contrast, heat treatment ht 2 generates a significant increase in the ϵ-phase. In addition, γ-phase is still detected. Table 2.2 summarizes the results of the ferromagnetic phase (α'-phase) content measurements. In the sht state, approximately 50 vol.% of the specimen consists of martensite. The first tempering treatment (ht 1) results in a slight increase of the amount of ferromagnetic phase. After the ht 2 heat treatment the amount of α'-phase reduced to 15 vol.%.

In Fig. 2.4 the microstructure of the alloy in the three different heat-treatment states (sht, ht 1 and ht 2) is presented. The alloys feature a complex microstructure which consists mainly of two domains. The areas that appear dark in the micrographs are assigned to ferromagnetic α'-martensite, whereas the bright areas consist of paramagnetic austenite and/or ϵ-martensite [22]. In the sht state (Fig. 2.4a) the alloy reveals a fine microstructure, but the insert image taken at a

Table 2.2: Results of the ferrite (α'-phase) content measurements (in vol.%).

Heat treatment state	Ferrite content
sht (1100 °/2 h + subsequently water quenched)	49 ± 1
ht 1 (sht + 500 °C/1 h)	56 ± 1
ht 2 (sht + 700 °C/10 min + 500 °C/10 h)	15 ± 1

2.4 Results

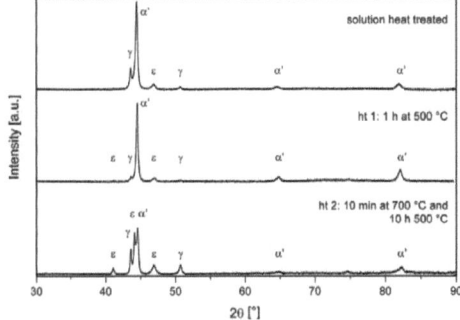

Figure 2.3: XRD spectra of Fe-10Mn-1Pd in three different heat treatment states, indicating the influence of heat treatments on the relative amount of the respective phases. In addition to the expected α'- and γ-phases, the hcp ϵ-phase is detected.

low magnification indicates segregations not leveled out during the forging and solution heat-treatment. Upon heat treatments (ht 1 and ht 2), the microstructure changes moderately. The bright areas assigned to the γ- and/or ϵ-phase are more connected and larger.

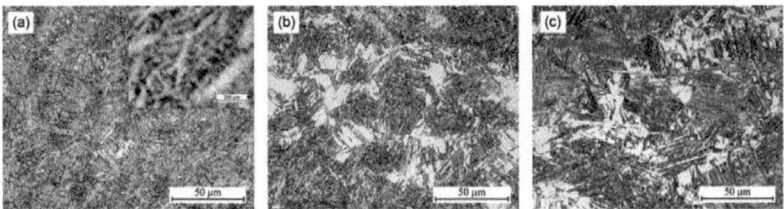

Figure 2.4: Optical micrographs of Fe-10Mn-1Pd specimens in three different states of heat-treatment: (a) solution heat-treated; the insert taken at a low magnification indicates segregations. (b) ht 1 state (sht + 1 h at 500 °C). (c) ht 2 state (sht + 10 min at 700 °C and subsequently aged for 10 h at 500 °C). In all heat treatment states, the samples feature a complex microstructure consisting of α'-martensite, retained austenite (γ-phase) and/or ϵ-martensite.

41

2 Design strategy

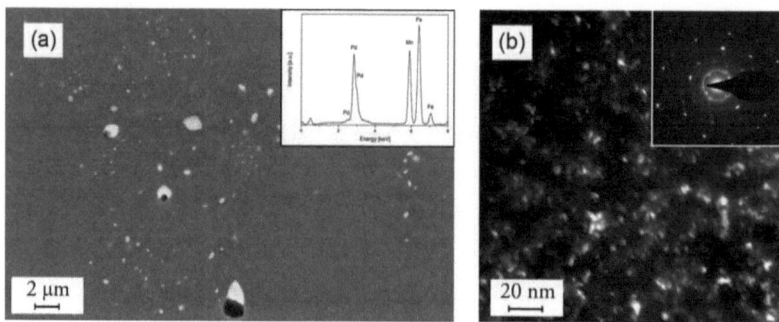

Figure 2.5: (a) SEM image (backscatter electron contrast) of Fe-10Mn-1Pd in the sht state. Embedded in the Fe–Mn matrix, two different particles can be observed: the IMPs appearing brightly are of type (Fe,Mn)Pd, and the dark particles are iron-manganese sulfide inclusions. The insert shows the EDX spectrum of a (Fe,Mn)Pd particle. (b) TEM dark field image of Fe-10Mn-1Pd (sht + 20 min at 550 °C) showing secondary Pd-rich IMPs formed during the annealing.

The SEM image in Fig. 2.5a shows a Fe-10Mn-1Pd alloy in the sht condition. The image was taken in the backscattered electron mode and shows chemical contrast in the sample. Embedded in the Fe–Mn matrix a number of brightly appearing particles is observed. Their size ranges from approximately 300 nm to 1 μm. EDX analysis performed in the SEM and TEM indicate that these particles are of the type (Fe,Mn)Pd, i.e. they are highly enriched in palladium as indicated in the insert in Fig. 2.5a showing the EDX spectrum taken from such a particle. According to EDX analysis, there is, however, still > 0.8 wt.% Pd remaining in the matrix. Furthermore, EDX analysis done on the particles that appear dark in the backscattered electron image indicates that they are iron-manganese sulfides.

In Fig. 2.5b a dark field TEM image of the microstructure of a sample in an aged state comparable to ht 1 is shown. It illustrates as an example the formation of secondary Pd-containing IMPs during aging treatments. Their size ranges from approximately 5 nm to 10 nm. The image is from an ongoing study characterizing the precipitation behavior of Fe–Mn–Pd alloys. For simplification reasons (in order to exclude the influence of carbon) the alloy was manufactured using pure iron instead of carbon steel, but the sample was produced in the same way as described in the experimental section. The heat treatment includes annealing for 20 min at 550 °C. Electron diffraction analysis (inset in Fig. 2.5b) revealed that these particles are of the type (Fe,Mn)Pd [26].

The results of the hardness measurements are given in Table 2.3. Pure iron (Armco) shows a low hardness value compared to the alloys containing Mn. In the sht condition the hardness values of the alloys (Fe, Fe-10Mn and Fe-10Mn-1Pd) are high, above 400 HV10. The alloys containing Mn are slightly harder than the carbon steel. The aging treatments subsequent to the sht condition generally lead to a decrease in the hardness values. The highest decay of the hardness value is measured consequent to ht 2.

2.4.2 Immersion testing and impedance spectroscopy

Immersion testing results are presented in Fig. 2.6a where the mass loss per area as a function of immersion time is presented. Carbon steel shows the lowest degradation rate, indicated by the slope of the curve. The Fe–Mn alloy (ht 2) shows an increased degradation rate compared to the carbon steel. The highest degradation rate, however, was found for Fe-10Mn-1Pd (here shown in ht 2 condition as example). These samples also show a fairly high standard deviation, which arises from the difficulty of removing all degradation products deposited on the sample surface during immersion.

The inset shows a SEM image (secondary electron image) of a Fe-10Mn-1Pd (ht 2) sample immersed for 48 h in SBF and subsequently cleaned as described above. The degradation products still partially cover the sample surface. EDX measurements indicate the presence of significant amounts of calcium and phosphorus additionally to Fe and Mn in this layer. Similar results for degradation deposition were also obtained by Hermawan et al. for Fe-35Mn [6]. The regions where the degradation products are removed reveal only signals from the matrix. The degradation apparently proceeds homogeneously and reveals a surface topography resembling that obtained by metallographic etching (Fig. 2.4).

The results of the EIS measurements in SBF are presented by Bode plots (Fig. 2.6b). The carbon steel reveals the highest polarization resistance (approximately

2 Design strategy

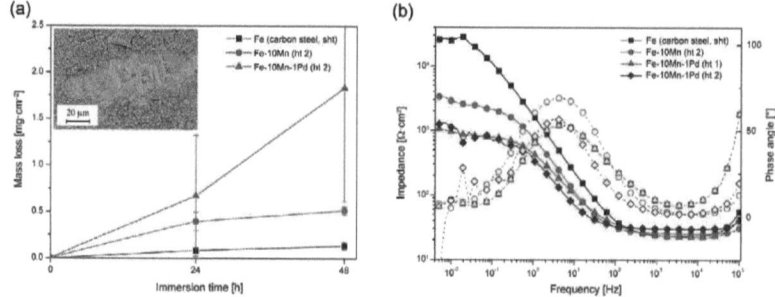

Figure 2.6: (a) Mass loss over immersion time of Fe (carbon steel, sht), Fe-10Mn (ht 2) and Fe-10Mn-1Pd (ht 2) subjected to immersion testing. The insert shows a SEM image (secondary electron contrast) of the surface of a Fe-10Mn-1Pd (ht 2) sample immersed for 48 h in SBF and cleaned as described. (b) Bode plots showing the impedance (solid symbols) and phase shift values (open symbols) of Fe (sht), Fe-10Mn (ht 2) and Fe-10Mn-1Pd in two different heat treatment states (ht 1 and ht 2).

26 kΩ·cm^2). Alloying with Mn results in a decrease of the polarization resistance to about 3.3 kΩ·cm^2. The alloys containing Pd feature even lower polarization resistances with lowest values in the aged conditions (ht 1: 1.0 kΩ·cm^2 and ht 2: 1.2 kΩ·cm^2).

2.4.3 Mechanical properties

The mechanical properties of the alloys evaluated are summarized in Table 2.3. A selection of the corresponding stress-strain curves is shown in Fig. 2.7. The values for pure iron (Armco quality) are illustrated for comparison. The carbon steel in the sht state shows reasonable strength values of about 900 MPa but poor uniform elongation of only 3.5%. The Fe–Mn(–Pd) alloys in the sht state are rather brittle and failed before reaching uniform elongation. Therefore no values for mechanical properties are given for Fe-10Mn (sht). The alloy Fe-10Mn-1Pd (sht) showed only a little plastic deformation before fracture.

The heat treatments ht 1 and ht 2 generate a significant increase in the ductility at high strength levels (Table 2.3). For the alloy Fe-10Mn-1Pd fairly good elonga-

2.5 Discussion

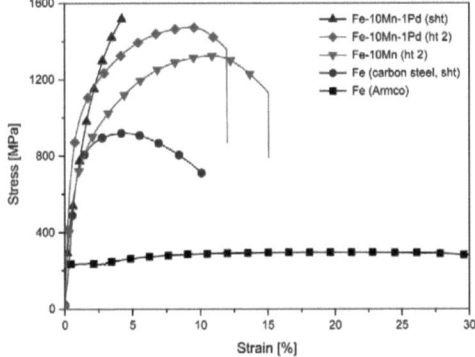

Figure 2.7: Tensile testing curves of as-received pure iron (Armco), carbon steel and Fe-10Mn-1Pd in the sht state; and Fe-10Mn and Fe-10Mn-1Pd in the ht 2 (tempered) state. Fe-10Mn-1Pd (ht 2) shows increased strength compared to Fe-10Mn (ht 2) due to strengthening by Pd-IMPs.

tion to fracture levels of $\geq 7\%$ are achieved, at a strength domain of ≥ 850 MPa (yield stress, YS) and ≥ 1450 MPa (ultimate tensile stress, UTS). The Pd-free alloy exhibits slightly higher ductility values, but at a significantly lower strength level; the increase in UTS caused by Pd-addition is about 150 MPa. In the ht 1 condition the yield and ultimate tensile strength of the alloys are slightly higher compared to the values measured in the ht 2 state whereas the uniform elongation and the elongation at fracture stay at comparable levels.

2.5 Discussion

2.5.1 Microstructure

The microstructural diversity of the binary Fe–Mn system has been described extensively in literature [22, 27, 28]. There are two main peculiarities that facilitate creation of complex structures: (i) the various martensitic transformations, and (ii) the pronounced transformation hysteresis. By cooling austenitic γ-Fe–Mn solid solutions three different martensitic transformations occur depending on

2 Design strategy

Table 2.3: Mechanical properties of the alloys at room temperature.

Designation	YS (MPa)	UTS (MPa)	ϵ_u (%)	ϵ_f (%)	H (HV10)
Fe (Armco)	250	300	19.5	37.5	85 ± 1
Fe (sht)	700	900	3.5	9.5	403 ± 2
Fe-10Mn (sht)	NA*	NA*	NA*	NA*	428 ± 6
Fe-10Mn (ht 1)	800	1400	9.5	14.0	384 ± 5
Fe-10Mn (ht 2)	650	1300	9.0	14.0	374 ± 7
Fe-10Mn-1Pd (sht)	950	1500	2.0	2.0	432 ± 8
Fe-10Mn-1Pd (ht 1)	900	1550	6.5	7.0	437 ± 3
Fe-10Mn-1Pd (ht 2)	850	1450	8.0	11.0	376 ± 6

YS, yield strength; UTS, ultimate tensile strength; ϵ_u, uniform elongation; ϵ_f, elongation at fracture; H, hardness

Due to the limited number of samples available, only approximate values are given here. * Samples failed before reaching uniform elongation.

the Mn content, namely $\gamma \rightarrow \alpha'$ in the concentration range between 3 and 12 wt.%, $\gamma \rightarrow \epsilon$ between 12 and 30 wt.% Mn and $\epsilon \rightarrow \alpha'$ in the range between 12 and 18 wt.%. The lines in Fig. 2.2 illustrate the start (M_s) and finish (M_f) temperatures of the martensitic transitions, as well as the temperatures for the reverse transition (A_s and A_f). According to Fig. 2.2 we would expect the microstructure of the sht alloys containing 10 wt.% Mn to consist mainly of the α'-phase. However, in the alloy Fe-10Mn-1Pd investigated in this study, the M_s and M_f temperatures (measured by means of dilatometry) are significantly decreased compared to the reference data, as indicated by the data points in Fig. 2.2. The decrease in Ms temperature can be attributed to the alloying with Pd and C, because these elements increase the stability of the γ-phase (see the well known Fe–C PD and the Fe–Pd PD in Fig. 2.1b). The As temperature agrees well with literature data, whereas A_f is shifted to higher temperatures. The increase in the transformation range can be attributed to the manganese segregation, in fact for both, the M_s to M_f and the A_s to A_f range.

2.5 Discussion

The altered transformation temperatures have an influence on the phase fractions present in the alloys. The results of the ferrite-content measurements state that the Fe–Mn–Pd alloy contains only 50 vol.% α'-phase. According to the results of the XRD measurements (Fig. 2.3) the remaining phases are paramagnetic and correspond to the γ-phase and/or ϵ-martensite. Such deviations from the "real" PD can also be attributed to the Mn segregations as seen in the optical micrographs in Fig. 2.4a. In regions of higher Mn content the γ-phase is stabilized, generating a decrease in the M_s and M_f transition temperatures, and the simultaneous formation of the ϵ-phase.

Surprisingly, already in the sht condition "primary" (Fe,Mn)Pd-IMPs are found in the alloys (Fig. 2.5a): the temperature for the solution heat treatment was chosen according to the information available in the binary PDs (Fe–Mn, Fe–Pd and Mn–Pd). According to these PDs, the alloy should consist of a single, fully austenitic phase at 1100 °C. Apparently, the solution heat treatment did not induce complete solubility of Pd in the austenitic Fe-Mn matrix. In contrast, annealing at 1250 °C resulted in a microstructure free of particles and is more likely a true solution heat treatment. Obviously, in the ternary Fe–Mn–Pd system the (Fe,Mn)Pd-IMPs possess a significantly higher stability than estimated from the binary systems. Nevertheless, it was adhered to the temperature of 1100 °C since the main goal of the design strategy is to increase the degradation rate by the formation of noble IMPs regardless of their formation history.

The heat treatment 1 induces a tempering of the α'-martensite present in the sht condition. Since the alloy is not heated above the A_s temperature the α'-martensite is not expected to transform into austenitic phase. The XRD-spectrum (Fig. 2.3) reveals that the relative amount of γ-phase is slightly decreased compared to the sht condition. Such behavior is well known from Fe–C martensite where the retained austenite partly transforms to martensite during tempering [29].

During ht 2 at 700 °C a diffusion-controlled transition to fully austenitic microstructure would be expected to occur. However, due to the high A_f temperature (> 700 °C) most probably an incomplete $\alpha' \rightarrow \gamma$ transformation will take place. Nevertheless a significant increase in γ-phase can be assumed. During the

2 Design strategy

subsequent isothermal heat treatment at 500 °C for 10 h the systems attempts to approach equilibrium conditions: the Fe–Mn PD (Fig. 2.1a) shows a two-phase field for the alloy Fe-10Mn at 500 °C. The segregations already present in the material actually progress further: the alloy tends to separate into a Mn-poor ferritic phase containing only about 3 wt.% Mn and a Mn-rich austenitic phase with about 20 wt.% Mn. The formation of Mn-poor ferritic regions in turn stabilizes the adjacent γ regions by the excess Mn. This leads to a considerable increase in the ϵ-martensite content after ht 2 compared to the sht condition (Table 2.2).

Simultaneously to the phase transitions, in the Pd-containing alloys Pd-rich (secondary) IMPs (Fe,Mn)Pd are precipitated (Fig. 2.5b) because of the reduced solubility of Pd in the Fe–Mn matrix at 500 °C. Due to the low diffusivity of the IMP-forming elements at this temperature the (Fe,Mn)Pd phase develops as very fine precipitates (< 10 nm). These secondary IMPs have an influence on the degradation performance and the mechanical properties as discussed in the following.

2.5.2 Influence of the microstructure on the degradation performance

Immersion testing and EIS were chosen to characterize the degradation performance of the newly developed alloys since these techniques are both straightforward and easy implementable. Even though they rely on different measuring techniques (weight loss versus impedance measurements), they yield a comparable ranking of the alloys regarding their degradation performance.

Fe (carbon steel) acts as reference concerning the degradation performance. As shown in Fig. 2.6, Fe–Mn alloys feature an increased degradation rate compared to iron, which was also shown by Hermawan et al. [5]. The according increase in the corrosion rate is attributed to the reduced standard electron potential, i.e. the less noble state of the Fe–Mn solid solution [19]. By alloying Pd the degradation rate is increased even further as expressed by the increase in the mass loss and the decrease in the polarization resistance. According to EIS measurements (Fig. 2.6b), the Fe–Mn–Pd alloys reveal a corrosion susceptibility one order of magnitude higher than observed for carbon steel.

The Fe–Mn–Pd alloys in the two different heat treatment states (ht 1 and ht 2) show comparable degradation performance. This indicates that both heat treatments generate a similar microstructural situation regarding the Pd-rich IMPs. The samples feature macroscopically homogeneous degradation behavior, which is attributed to the formation of homogeneously distributed IMPs during solution heat treatment as well as during the subsequent aging heat treatments.

As mentioned earlier, the pronounced adherence of the degradation products on the sample surface is responsible for the high standard deviation of the immersion testing measurements. This makes the interpretation of these results demanding and for future experiments, alternative measuring techniques have to be considered, as e.g. the determination of the ion release by means of inductive coupled plasma – mass spectrometry (ICP-MS).

2.5.3 Influence of the microstructure on the mechanical performance

Pure iron (Armco) shows only low strength but features high ductility. However, for potential applications, higher strength levels are desirable. Adding of C to Fe has a tremendous influence on mechanical properties: strength increases to about 900 MPa and ductility decreases significantly. Fe–Mn steels exhibit even higher strength levels. Their martensite-rich microstructure reveals high hardness but low ductility in the sht condition (Fig. 2.7 and Table 2.3). The brittleness of the Fe–Mn(–Pd) alloys is attributed to C in the α'-phase in SSSS. In the γ-phase, Fe and C form a solid solution. Upon quenching diffusion of C is suppressed and a sudden reorientation from the austenitic γ-Fe to a body-centered tetragonal solid solution (martensite) occurs. The complex crystal structure and the supersaturated concentration of C generate the characteristic brittle nature.

Generally, martensitic steels must be tempered to increase their ductility [30]. As exemplarily shown in Table 2.3 and Fig. 2.7, the heat treatments evaluated in this study have a significant influence on the alloy's mechanical properties. Compared to the sht state increased ductility at high strength can be achieved by both heat treatments (ht 1 and ht 2). The complex nature of the microstructure makes it difficult; however, to elucidate the contributions of the different phases to the overall strength of the material and a detailed explanation will be published elsewhere.

Compared to Fe–Mn alloys, the Fe–Mn–Pd samples in the same heat treatment state show increased strength values: The yield strength of Fe-10Mn-1Pd (ht 1) is about 100 MPa higher than that of Fe-10Mn (ht 1). Similarly, the yield strength of Fe-10Mn-1Pd (ht 2) is approximately 200 MPa higher than that of Fe-10Mn (ht 2). Obviously, the increase in strength is generated by the secondary Pd-IMPs according to the particle hardening mechanism. Like in other particle hardening systems (e.g. hardening Al-alloys) the mechanical properties can be adjusted in a wide range by varying the heat treatment parameters (i.e. temperature and time of heat treatment).

2.5.4 Efficiency and potential of the design strategy

The alloy Fe-10Mn-1Pd exemplarily emphasizes the efficiency of the design strategy presented in this study. Compared to the Fe–Mn system or to pure iron the newly-developed Fe-10Mn-1Pd alloy exhibits a significantly increased degradation rate, which is, to some extent, adjustable by appropriate heat treatment parameters. In addition, this alloys feature attractive mechanical properties which also rely on the heat treatment state. Although possessing an interesting profile, the alloy Fe-10Mn-1Pd is not ideal for certain applications because of the presence of the ferromagnetic α'-phase. This deficiency can be overcome, however, by systematically varying the Mn-content and heat treatment parameters to adjust the amount of each phase in the alloys: low Mn-contents result in a large amount of ferromagnetic α'-phase while the increased Mn-contents stabilize the austenitic phase (Fig. 2.1a). Upon quenching from elevated temperatures the austenitic γ-phase martensitically transforms into ϵ-phase. Both phases, γ and ϵ, are paramagnetic. According to the "real" transformation diagram in Fig. 2.2, alloys containing more than approximately 18 wt.% Mn are free of α'-phase. This is emphasized by ongoing evaluations of Fe–Mn–Pd alloys containing 20 wt.% Mn. Here, the formation of the ferromagnetic α'-phase is completely suppressed which not only has beneficial influence on the mechanical properties, but also qualifies such alloys for applications where MRI (magnetic resonance imaging) is required.

Upcoming studies will cover detailed evaluation of the influence of chemical composition and heat treatment parameters on the microstructure and the corresponding electrochemical and mechanical properties of Fe–Mn–Pd alloys. Preliminary results of alloys containing 15 and 20 wt.% Mn indicated the validity of the design strategy also for such alloys and emphasize the design approach. The potential of the design strategy is considered high given the high degree of freedom it provides. The variation of the alloy's Mn-content and the manifold possibilities of different heat treatments result in a great spectrum of electrochemical and mechanical properties that can be specifically tailored according to the needs of applications.

2.6 Conclusions

In this study, a design strategy for the development of biodegradable Fe-based alloys for medical applications is presented. The goal of the strategy was to achieve an alloy's performance appropriate for degradable implant applications including both, enhanced degradation rate compared to pure iron, and suitable strength and ductility. The design approach relies on the controlled modification of the microstructure of iron by suitable alloying and appropriate heat treatment parameters. The alloying elements have the function of decreasing the degradation resistance of the iron taking into account two aspects: (i) the formation of a solid solution such that the Fe-matrix is turned more susceptible to corrosion, and (ii) the formation of noble IMP particles that generate micro-galvanic corrosion and promote active dissolution of the matrix. The IMP particles are also deployed in strengthening the alloy by particle hardening. Manganese and palladium have shown to be suitable alloying additions for this approach. The heat treatment parameters are chosen such that the efficacy of the IMP particles is enhanced (by restricting their size and generating a homogeneous distribution). In addition, they are also used to adjust the phase content of the alloys, which in turn has an impact on the mechanical properties. With regard to the microstructural characteristics of the newly-developed Fe-10Mn-1Pd alloy (in wt.%) resulting in enhanced degradation rate and attractive mechanical performance, our design strategy demonstrates high efficiency. It also offers high degree of metallurgical freedom and allows tailoring properties of Fe–Mn–Pd alloys according to the requirements of specific degradable implant applications. In upcoming studies this is elaborated in detail by evaluating the influence of chemical composition and heat treatment parameters on the alloy's performance.

Acknowledgements

The authors thank A.S. Sologubenko, F.H. Dalla Torre and F. Moszner for performing TEM analysis, C. Pecnik for the support in immersion testing and EIS measurements, R. Zoller for help with the microstructure analysis, and E. Fischer for help with the sample production. This research was funded by the project "BioCompatible Materials and Applications – BCMA" initiated by the AIT Austrian Institute of Technology GmbH and by the Staub/Kaiser foundation (Switzerland).

References

[1] Staiger MP, Pietak AM, Huadmai J, Dias G. Magnesium and its alloys as orthopedic biomaterials: A review. Biomaterials 2006;27:1728-34.

[2] Mani G, Feldman MD, Patel D, Agrawal CM. Coronary stents: A materials perspective. Biomaterials 2007;28:1689-710.

[3] O'Brien B, Carroll W. The evolution of cardiovascular stent materials and surfaces in response to clinical drivers: A review. Acta Biomater 2009;5:945-58.

References

[4] Hänzi AC, Gunde P, Schinhammer M, Uggowitzer PJ. On the biodegradation performance of an Mg–Y–RE alloy with various surface conditions in simulated body fluid. Acta Biomater 2009;5:162-71.

[5] Hermawan H, Moravej M, Dubé D, Fiset M, Mantovani D. Degradation Behaviour of Metallic Biomaterials for Degradable Stents. Adv Mater Res 2007;15-17:113-8 [THERMEC 2006 Supplement].

[6] Hermawan H, Dubé D, Mantovani D. Development of Degradable Fe-35Mn Alloy for Biomedical Application. Adv Mater Res 2007;15-17:107-12 [THERMEC 2006 Supplement].

[7] Hermawan H, Alamdari H, Mantovani D, Dube D. Iron-manganese: new class of metallic degradable biomaterials prepared by powder metallurgy. Powder Metall 2008;51:38-45.

[8] Witte F, Kaese V, Haferkamp H, Switzer E, Meyer-Lindenberg A, Wirth CJ, Windhagen H. In vivo corrosion of four magnesium alloys and the associated bone response. Biomaterials 2005;26:3557-63.

[9] Witte F, et al. In vitro and in vivo corrosion measurements of magnesium alloys. Biomaterials 2006;27:1013-8.

[10] Zartner P, Buettner M, Singer H, Sigler M. First biodegradable metal stent in a child with congenital heart disease: Evaluation of macro and histopathology. Cathet Cardiovasc Interv 2007;69:443-6.

[11] Erbel R, et al. Temporary scaffolding of coronary arteries with bioabsorbable magnesium stents: a prospective, non-randomised multicentre trial. Lancet 2007;369:1869-75.

[12] Peuster M, Hesse C, Schloo T, Fink C, Beerbaum P, von Schnakenburg C. Long-term biocompatibility of a corrodible peripheral iron stent in the porcine descending aorta. Biomaterials 2006;27:4955-62.

[13] Mueller PP, May T, Perz A, Hauser H, Peuster M. Control of smooth muscle cell proliferation by ferrous iron. Biomaterials 2006;27:2193-200.

[14] Zhu S, Huang N, Xu L, Zhang Y, Liu H, Sun H, Leng Y. Biocompatibility of pure iron: In vitro assessment of degradation kinetics and cytotoxicity on endothelial cells. Mater Sci Eng C 2009;29:1589-92.

[15] Serruys PW, Kutryk MJB, Ong ATL. Drug therapy - Coronary-artery stents. New Engl J Med 2006;354:483-95.

[16] Vargel C, Jacques M, Schmidt MP. Corrosion of Aluminium: Elsevier, 2004.

[17] Revie WR, Uhlig HH. Corrosion and Corrosion Control: John Wiley & Sons, 2008.
[18] Predel B. Landolt-Börnstein. In: Madelung O, editor. Group IV - Physical Chemistry, vol. 5e. Springer-Verlag, 1998.
[19] Kawashima A, Asami K, Hashimoto K. Effect of Manganese on the Corrosion Behaviour of Chromium-Bearing Amorphous Metal-Metalloid Alloys. Sci Rep Res Inst Tohoku Univ Phys Chem Metall 1981;29:276-83.
[20] Wataha JC, Hanks CT. Biological effects of palladium and risk of using palladium in dental casting alloys. J Oral Rehabil 1996;23:309-20.
[21] Geurtsen W. Biocompatibility of dental casting alloys. Crit Rev Oral Biol Med 2002;13:71-84.
[22] Schumann H. Die martensitischen Umwandlungen in kohlenstoffarmen Manganstählen. Arch Eisenhüttenw 1967;38:647-56.
[23] Ardell AJ. Precipitation Hardening. Metall Trans Phys Metall Mater 1985;16:2131-65.
[24] Müller L, Müller FA. Preparation of SBF with different HCO_3^- content and its influence on the composition of biomimetic apatites. Acta Biomaterialia 2006;2:181-9.
[25] Scully JR. Polarization resistance method for determination of instantaneous corrosion rates. Corrosion 2000;56:199-218.
[26] PANalytical. X'Pert HighScore Plus 2.1.0. Almelo, The Netherlands, 2004.
[27] Cotes S, Sade M, Guillermet AF. Fcc/Hcp Martensitic-Transformation in the Fe–Mn System - Experimental-Study and Thermodynamic Analysis of Phase-Stability. Metall Trans Phys Metall Mater 1995;26:1957-69.
[28] Jun J-H, Choi C-S. The influence of Mn content on microstructure and damping capacity in Fe-(17~23)%Mn alloys. Mater Sci Eng 1998;252:133-8.
[29] Speich GR. Metals Handbook. Metals Park, Ohio: American Society of Metals, 1973.
[30] Krauss G. Steels: Processing, Structure, and Performance: ASM International, 2005.

3 Microstructure and mechanical performance

In the Chapter 2 the Fe-10Mn-1Pd alloy was presented as an example of a high-strength material put together according to the design strategy. High-manganese and high-carbon steels offer significantly higher ductility levels and a pronounced strain-hardening behavior. These characteristics make these materials interesting from the application point of view, and the design strategy was extended to cover them. Chapter 3 presents a thermomechanical treatment for optimizing their microstructure and mechanical performance such that they may be deployed as degradable implant material.

Recrystallization behavior, microstructure evolution and mechanical properties of biodegradable Fe–Mn–C(–Pd) TWIP alloys[1]

In this study the interplay between recrystallization and precipitation in a biodegradable TWIP steel developed for the use in temporary implants was investigated. Microstructural and mechanical properties were studied and a thermomechanical treatment was designed with the aim of achieving an overall performance suitable for the intended application as temporary implant material. The formation of Pd-rich precipitates in the cold-worked state was found to considerably retard recrystallization during an annealing treatment. The formation, morphology and interaction with dislocations of these precipitates were studied by means of scanning and transmission electron microscopy. Grain boundary pinning by Pd-rich precipitates (Zener drag) and reduced dislocation mobility due to a solute drag effect caused by the enrichment of dislocation cores with Pd were both identified as mechanisms which impede recrystallization. A model is reported which explains the interplay between recrystallization and precipitation, and provides the basis for the optimized thermo-mechanical treatment then presented. The resulting mechanical properties, in particular the combination of high strength and ductility with a pronounced strain hardening response, exceeds the performance of other TWIP steels and alloys typically used in biomedical implants, such as stainless steel, titanium or cobalt-chromium alloys. The specific property profile developed is especially advantageous for the production and deployment of cardiovascular stents.

3.1 Introduction

Because of its favorable combination of mechanical, biological and electrochemical properties iron is an interesting candidate for temporary implants, as has

---

[1]M. Schinhammer, C.M. Pecnik, F. Rechberger, A.C. Hänzi, J.F. Löffler, P.J. Uggowitzer; Acta Materialia 60 (2012) 2746-56

been shown in various in vitro and in vivo studies [1-5]. For complex applications such as cardiovascular stents, the material's overall mechanical performance plays a key role. Deploying high-strength alloys makes possible the design of filigree implants, which reduce the amount of material taken up by the body, and high ductility facilitates manufacturing and stent deployment, rendering its application safe. Pronounced strain hardening supports the uniform dilatation of the stent further and therefore reduces physical irritations. Since the degradation rate and especially the mechanical properties of pure iron itself are not appropriate for stent materials [3, 6], a design strategy for biodegradable Fe–Mn–Pd alloys was developed with the aim of improving the performance of Fe-based alloys for such applications [7]. Its goals are an increased degradation rate and mechanical performance superior to that of pure iron. Adding Mn within the solubility limit of Fe results in a decrease of the standard electrode potential of the matrix, making it less noble and therefore more susceptible to corrosion. Palladium was added in small amounts to form small and homogeneously distributed Pd-rich precipitates which cause microgalvanic corrosion [7, 8].

In addition to their influence on electrochemical properties, alloying elements are used to modify the mechanical performance. Depending on the Mn content, a variety of different microstructures (consisting of α- and/or ϵ-martensite, austenite) can be obtained in the Fe–Mn system [9]. This influences the deformation modes (dislocation slip, strain-induced phase transformations, twinning) and consequently also the mechanical properties of the alloys [10, 11]. Prominent Fe–Mn alloy families are maraging steels [12], TRIP (TRansformation Induced Plasticity) and TWIP (TWinning Induced Plasticity) steels [13-17]. These alloys combine moderate yield strengths (typically of 350 MPa), high ultimate tensile strengths (800 – 1000 MPa), pronounced work-hardening capacity and high ductility (elongation to fracture of up to 100%) [13]. TWIP steels, which are the focus of this study, contain a high Mn content (commonly between 18 and 30 wt%) and minor amounts of further alloying elements such as C, Si, Al and N [14-16, 18-20]. The concentrations of alloying elements are chosen so as to fully stabilize the austenitic phase (also upon mechanical deformation) and to adjust the stacking fault energy (SFE) to values between 20 and 40 mJ·m^{-2} [14]. The high strain hardening capacity of TWIP alloys is commonly attributed to the reduced mean free path of dislocations resulting from the creation of deformation twins, which act as strong obstacles to dislocation glide [14]. As it has previously been shown for martensitic Fe–Mn–Pd alloys, adding Pd (typically 1 wt%) generates Pd-rich precipitates which increase the degradation rate and influence the mechanical properties, i.e. increase the strength of the material [7, 8]. Moszner et al. [8] demonstrated that the Pd-rich precipitates tend to form along dislocation lines via pipe diffusion, as is also characteristic of maraging steels [21]. Because the alloys investigated in this study have a stable austenitic microstructure, a comparison with microalloyed steels is more appropriate. Elements such as Nb, Ti or V are added to the latter in order to retard austenite recrystallization during hot deformation via solute drag and by producing dispersed pinning particles [22-25]. The strain-induced precipitation of these particles occurs mainly on dislocation structures in the matrix [22, 23, 26], and a high dislocation density accelerates the formation of the precipitates, again assisted by pipe diffusion [24, 27]. Considering the rapid formation of the Pd-rich precipitates in the martensitic Fe–Mn–Pd alloys, it is likely that in the alloys of this study the added Pd significantly influences the recrystallization behavior. We thus investigate the recrystallization and precipitation behavior of Pd-alloyed TWIP steels and the resulting microstructures. It is the aim of this study to understand their influence on the mechanical properties in order to produce Fe-based alloys suitable for temporary implant applications.

3.2 Experimental

3.2.1 Alloy preparation and thermo-mechanical treatments

Two alloys of nominal compositions Fe-21Mn-0.7C (labeled TWIP) and Fe-21Mn-0.7C-1Pd (in wt%, labeled TWIP-1Pd) were produced from pure elements and cast in a vacuum induction furnace under argon atmosphere. The Mn and Pd contents were measured using X-ray fluorescence spectroscopy. The carbon con-

3.2 Experimental

Table 3.1: Composition (in wt.%) of the alloys investigated.

Alloy	Fe	Mn	C	Pd
Fe-21Mn-0.7C	Bal.	20.9	0.69	—
Fe-21Mn-0.7C-1Pd	Bal.	21.2	0.71	1.1

centration was determined using a LECO carbon determinator. Table 3.1 summarizes the alloy compositions. The cylindrical ingots (diameter 75 mm) were forged at approximately 1000 °C to a diameter of 12 mm. The alloys were then solution-heat-treated (sht) for 10 h at 1250 °C under argon atmosphere and subsequently water-quenched. This state served as basis for the subsequent three steps, which are summarized in Fig. 3.1.

In a first step (step A), the interplay between recrystallization and precipitation was investigated. To achieve this, the alloys (i.e. TWIP and TWIP-1Pd) were swaged to a diameter of 8 mm, corresponding to a cold working degree (CWD) of 56%, and subsequently annealed for 10 min at temperatures between 500 and 1250 °C.

In a second step (step B), the influence of the degree of cold working on the recrystallisation behavior of the TWIP-1Pd alloy was investigated. Starting again from sht samples (12 mm in diameter), they were swaged, corresponding to CWD of 12%, 23%, 30% and 44%, and subsequently annealed for 30 min at temperatures between 700 and 900 °C. Note that these two steps were performed independently and both started from the sht state.

The aim was to develop a microstructure exhibiting favorable mechanical properties. Consequently, the results of the first two steps were combined and a thermo-mechanical treatment was devised based on these findings (step C). In the sht state, the grain size was found to be very large and inhomogeneous because of the high temperature (1250 °C) required for this heat treatment [8]. Therefore, based on the results of the first step, the TWIP-1Pd alloy was first recrystallized (rexx), i.e. subjected to 56% of cold working and annealed for 10 min at 1150 °C. Then, based on the results of the second step, it was again swaged, corresponding to 12% and 23% of cold working, and finally annealed

3 Microstructure and mechanical performance

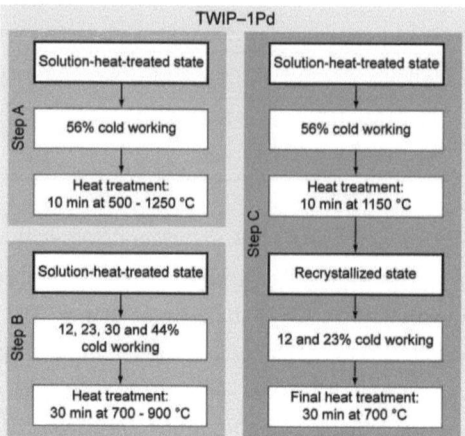

Figure 3.1: Flowchart illustrating the processing steps explored for TWIP–1Pd. Based on step A and B, an optimized thermo-mechanical treatment was developed (step C).

for 30 min at 700 °C. These materials are designated TWIP-1Pd CW12-ht and TWIP-1Pd CW23-ht. The specific parameters for this procedure are discussed in Section 3.4.1.3.

To determine the mechanical properties of the (Pd-free) TWIP alloy it was first cold-worked corresponding to a CWD of 30% and subsequently annealed for 30 min at 900 °C to fully recrystallize the material.

3.2.2 Microstructure characterization

The microstructure of the samples was analyzed via scanning electron microscopy (SEM), employing electron channeling contrast imaging (ECCI) using the backscattered electron (BSE) detector and by electron backscatter diffraction (EBSD). SEM studies were performed on a Hitachi SU-70 (Schottky-type field emission gun) equipped with a Nordlys EBSD camera and an X-max energy dispersive X-ray (EDX) detector (Oxford Instruments). The ECCI was performed using 10 kV acceleration voltage at 10 mm working distance. The EBSD scans used 20 kV acceleration voltage and a probe current of approximately 2 nA at 17 mm working distance. The samples were embedded in bakelite and ground as well as polished down to 0.25 μm diamond size, and the final polishing step was carried out using colloidal silica suspension (Buehler MasterMet 2). The indexing rate for the plots shown was generally above 93%, except for those of Fig. 3.3a (71%) and Fig. 3.6b (88%), due to the high amount of remaining deformed matrix and the large number of deformation twins, respectively.

Transmission electron microscopy (TEM) was performed on a FEI Tecnai F30 machine operated at 300 kV. Conventional TEM (CTEM) was deployed to assess the morphology of the microstructure and for defect structure analysis. The dark-field (DF) and atomic-number-sensitive high-angle-annular dark-field (HAADF) imaging modes of the scanning TEM (STEM) were used to show the defect structure and the compositional contrast, respectively. The TEM specimens were prepared by mechanically grinding the Fe-based samples to a thickness of approximately 100 μm. From these specimens, disks of 3 mm in diameter were punched out. These disks were then dimpled on both sides using a Gatan dimple grinder with 1 μm diamond suspension (Metadi oil-based). Electron transparency was obtained by twin-jet electro polishing (Tenu-Pol 5, Struers) using 24 V DC at a temperature of –30 °C. A solution of 5 vol.% perchloric acid in methanol served as the electrolyte.

The hardness (Vickers units HV10, 6 s indentation time) was determined using a Brickers 220 hardness measurement device (Gnehm, Switzerland). The mechanical properties were assessed by standard tensile testing (Schenck-Trebel, Germany) using round specimens of 3 mm diameter with a gauge length of 18 mm. The tests were performed using an initial strain rate of 10^{-3} s^{-1}.

3.3 Results

3.3.1 Influence of the annealing temperature on recrystallization behavior

The results of the hardness and grain size measurements after isochronal annealing of the 56% cold-worked samples are shown in Fig. 3.2a. The corresponding microstructures (EBSD and ECCI images) are given in Figs. 3.3 and 3.4. The ini-

3 Microstructure and mechanical performance

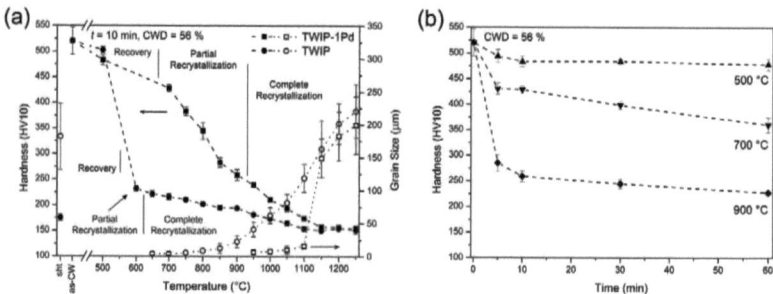

Figure 3.2: (a) Hardness and grain size as a function of heat treatment temperature after annealing of 56% cold-worked TWIP and TWIP-1Pd samples for 10 min. For the alloy TWIP-1Pd, four different regions can be distinguished. Solid symbols represent hardness values, whereas open symbols refer to the grain sizes measured. (b) Hardness evolution as a function of the annealing time for alloy TWIP-1Pd.

tially high hardness in the as-cold-worked state drops with increasing annealing temperature. For the TWIP-1Pd alloy, four different regimes can be distinguished. (i) In the range between 500 and 650 °C, only recovery occurs and the hardness slightly decreases. (ii) From 700 to 900 °C the hardness rapidly drops with increasing annealing temperature. Partial recrystallization takes place in this regime and a higher temperature results in a higher degree of recrystallization and hence a lower hardness. Concurrently, precipitation of Pd-rich particles occurs. (iii) Between 950 and 1100 °C the alloy completely recrystallizes and the grain size measured after annealing ranges from 6 to 16 μm. The formation of Pd-rich precipitates is also observed (see below). (iv) From 1150 to 1250 °C the alloy shows complete recrystallization without the occurrence of Pd-rich precipitates. The grain sizes in this temperature range are approximately 150 to 200 μm.

In contrast, the TWIP alloy shows complete recrystallization within 10 min of annealing at temperatures above 650 °C. Increasing annealing temperatures result in a steady decrease in the hardness and an increase in the grain size. Between 650 and 950 °C the grain sizes are between 4 and 40 μm. At higher temperatures, the grain size considerably increases.

3.3 Results

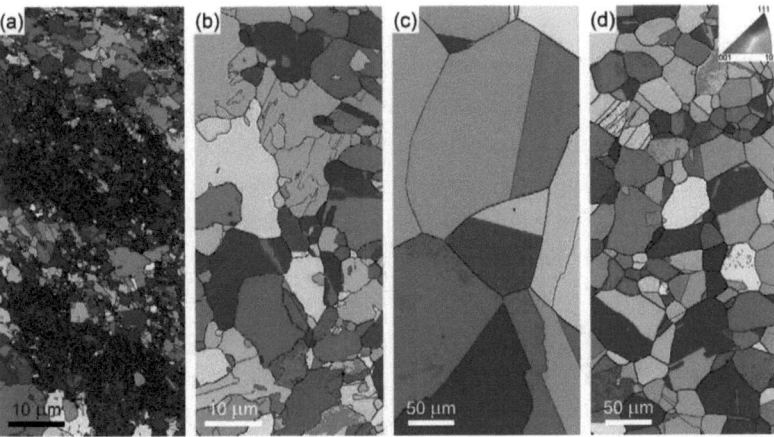

Figure 3.3: EBSD maps taken on the 56% cold-worked TWIP-1Pd (a-c) and TWIP (d) samples after four different annealing treatments. (a) Partial recrystallization takes place after 10 min at 700 °C. Fine grains which nucleated in a band-like arrangement and deformed matrix (which did not recrystallize) can be observed. High-angle grain boundaries are indicated by solid black lines and solid red lines correspond to 60° ⟨1 1 1⟩-twin boundaries. (b) The sample completely recrystallized after 10 min at 950 °C. (c) After annealing for 10 min at 1150 °C the sample shows complete recrystallization and a large grain size. (d) The microstructure of a (Pd-free) TWIP sample after annealing for 10 min at 950 °C. The colors correspond to the inverse-pole-figure (IPF) color coding in the out-of-plane direction (see inset).

Fig. 3.2b shows the hardness evolution as a function of annealing time for the TWIP-1Pd alloy for three different temperatures. The slight hardness drop upon annealing at 500 °C is attributed to recovery, as no recrystallized grains were found in these samples. During annealing at 700 °C, the measurements show an initial hardness decrease during the first 10 min due to partial recrystallization. The higher degree of recrystallization during annealing at 900 °C causes a more pronounced hardness drop, as indicated in the corresponding curve. Again, the most significant changes occur during the first 10 °min of annealing.

The EBSD maps of four different microstructures are shown in Fig. 3.3 and the corresponding ECCI images are given in Fig. 3.4. The images in Figs. 3.3a and 3.4a illustrate the microstructure of a 56% CW TWIP-1Pd sample after annealing for 10 min at 700 °C. The EBSD map shows a large number of nucleated grains which are arranged in a band-like structure. Areas of deformed matrix also still remain. The SEM images show a combination of orientation and atomic number contrast. The bright features (visible especially in the inset to Fig. 3.4a) are Pd-rich precipitates. EDX measurements confirmed the enrichment of Pd in the precipitates. As Pd is the heaviest constituent of the alloy, the presence of Pd generates a bright contrast in the BSE image. The precipitates are present in both the deformed matrix and the recrystallized grains. However, they are small in the former and larger ($\approx$20 to 50 nm) and thus better visible in the latter.

Upon annealing between 950 and 1100 °C the TWIP-1Pd samples show complete recrystallization, as indicated in Figs. 3.3b and 3.4b, which show the microstructure of a sample annealed for 10 min at 950 °C. The SEM images show that the grain boundaries of the recrystallized grains are decorated with Pd-rich precipitates, which are about 50 to 200 nm in size. Additionally, as indicated in the inset, a large number of small ($\approx$20 to 50 nm in size) precipitates are found within the grains.

The annealing at higher temperatures (1150 to 1250 °C) results in complete recrystallization without the occurrence of Pd-rich precipitates. Figs. 3.3c and 3.4c show the microstructure of a TWIP-1Pd sample annealed for 10 min at 1150 °C. The relatively high heat treatment temperature produces a large grain size, without Pd-rich precipitates (Fig. 3.4c). Even at higher magnification, as shown in the inset in Fig. 3.4c, none were detected.

The microstructure of a TWIP sample annealed for 10 min at 950 °C is shown in Figs. 3.3d and 3.4d. The EBSD map shows completely recrystallized and equiaxed grains which contain some annealing twins. The grain size of the TWIP sample is approximately 40 μm. This is considerably larger than that of the corresponding TWIP-1Pd sample (7 μm, c.f. also Fig. 3.2a), indicating the significant influence of the Pd-rich precipitates on the recrystallization process.

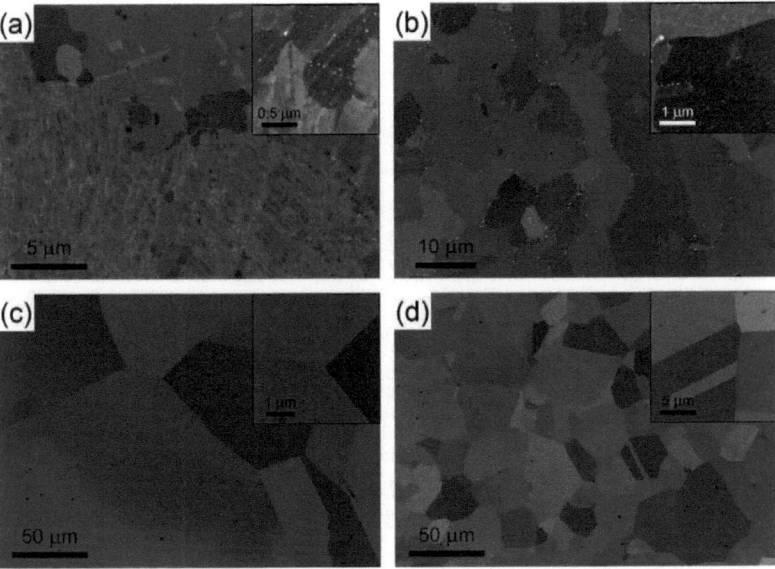

Figure 3.4: SEM images (ECCI) of the 56% cold-worked TWIP-1Pd (a-c) and TWIP (d) samples annealed at four different temperatures for 10 min. The insets show the same samples at higher magnifications. Note that the magnifications of the images were all selected according to their microstructure. (a) Partially recrystallized sample after 10 min at 700 °C. Pd-rich precipitates are present in the recrystallized grains and the deformed matrix. (b) After annealing for 10 min at 950 °C the sample completely recrystallizes and features Pd-rich precipitates at the grain boundaries. A large number of smaller precipitates are also present within the grains as indicated in the inset. (c) Annealing for 10 min at 1150 °C leads to complete recrystallization without precipitation of Pd-rich particles. (d) The (Pd-free) TWIP alloy annealed for 10 min at 950 °C exhibits a fully recrystallized microstructure.

3.3.2 Influence of the degree of cold working on recrystallization behavior

The influence of the degree of cold working and the annealing temperature on the recrystallization behavior is summarized in Fig. 3.5. During heat treatment at 500 °C only recovery occurs, which is consistent with the results presented in Fig. 3.2. At higher temperatures of 700 to 900 °C, the samples partially recrystallize, depending on the CWD. The threshold for the initiation of recrystallization during annealing at 700 °C lies above a degree of 23% cold working, while the increase of annealing temperatures shifts the threshold to lower CWD, as indicated by the dashed line. Accordingly, samples subjected to 12% cold working showed partial recrystallization during annealing at 900 °C.

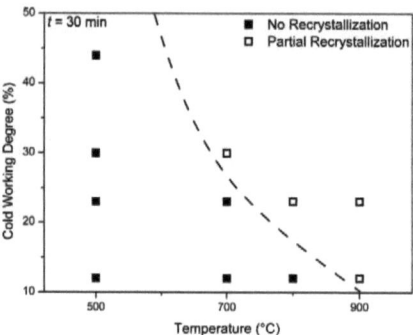

Figure 3.5: Influence of annealing temperature and degree of cold working on the recrystallization behavior of the TWIP-1Pd alloy. Recrystallization starts only above a critical temperature and critical degree of cold working.

3.3.3 Thermo-mechanical optimization

3.3.3.1 Microstructure characterization

The EBSD maps of the microstructure after the final heat treatment (annealing for 30 min at 700 °C) are shown in Fig. 3.6. The average grain size is around 150 μm and some annealing twins originating from recrystallization can be observed.

While the sample 12% cold-worked (Fig. 3.6a) shows an only moderate density of deformation twins, the sample deformed to a CWD of 23% shows a much higher density of such twins, as indicated in Fig. 3.6b. The insets depict an area of the map at higher magnification, i.e. measured with a finer step size. In neither case the onset of recrystallization could be detected.

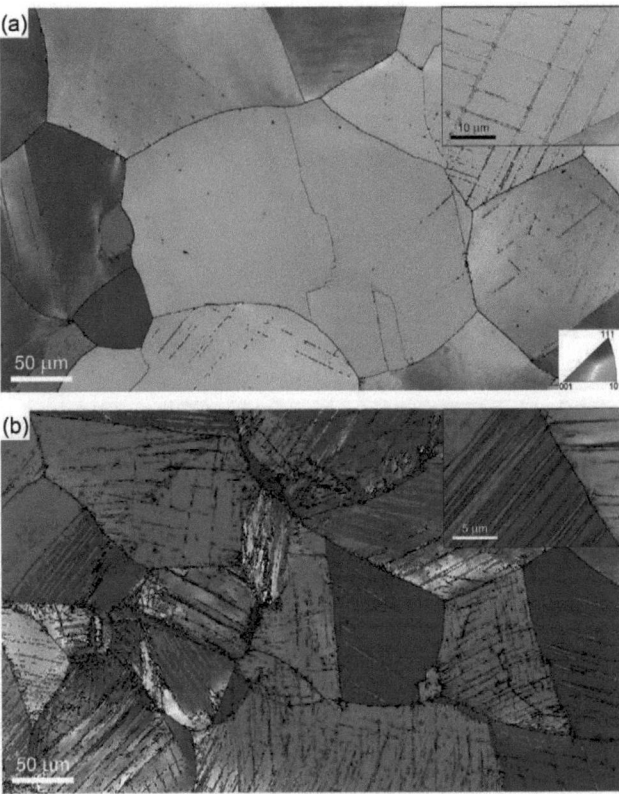

Figure 3.6: The EBSD maps (IPF out-of-plane color coding) show TWIP-1Pd samples in the final heat treatment state; cf. Fig. 3.1 and Section 3.2.1. (a) A 12% cold-worked sample annealed for 30 min at 700 °C (TWIP-1Pd CW12-ht). (b) A 23% cold-worked sample annealed for 30 min at 700 °C (TWIP-1Pd CW23-ht). The insets show an area of the samples with higher magnification. High-angle grain boundaries are indicated by solid black lines, and solid red lines correspond to 60° ⟨1 1 1⟩-twin boundaries.

The images in Fig. 3.7 show the TEM micrographs corresponding to the samples depicted in Fig. 3.6. The microstructure of a sample cold-worked to 23% and annealed for 30 min at 700 °C (Fig. 3.7a) features an array of deformation twins running diagonally across the image. The selected area diffraction pattern (see inset) acquired in the [0 1 1] zone-axis orientation from a larger area shows the corresponding reflections of the matrix and the twin. The deformation twins occur in arrays of twinned lamellae of about 20 to 50 nm in width. The matrix exhibits a high density of dislocations: these consist of short segments that are jagged and appear entangled. The corresponding dark-field STEM image (Fig. 3.7b) specimen shows small islands of Moirée contrast (indicated by the short black arrows). The islands are about 4 nm in size and their fringes are oriented in a small number of directions within the same grain. The latter fact indicates a definite crystallographic correspondence between the two lattices responsible for the interference. These Moirée islands (signified by arrows in Fig. 3.7b) are aligned along a dislocation line not clearly delineated under the illumination conditions employed. The image in Fig. 3.7c presents the microstructure of a sample which was 12% cold-worked and annealed for 30 min at 900 °C. The micrograph, acquired in a [1 1 2] zone-axis orientation, shows a large number of precipitates formed during annealing (indicated by long black arrows). They are predominantly elongated in shape and are about 20 to 50 nm in size. The precipitates were found within the twinned lamellae and at the twin boundaries. Under [1 1 2] zone axis illumination conditions, the precipitates cause no additional diffracted intensity modulations in the corresponding electron diffraction patterns (see inset). It is clearly seen that the dislocations interact with the precipitates. The atomic-number-sensitive HAADF-STEM image taken from about the same area is presented in Fig. 3.7d. The bright features marked by long black arrows are the precipitates seen in Fig. 3.7c. The precipitates appear bright in the HAADF STEM micrograph because they are enriched with the alloy's heaviest constituent Pd. Line-like bright features are also visible in the image (indicated by short white arrows). These thin short lines are either straight or curved and are due to dislocation lines enriched with Pd atoms.

3.3 Results

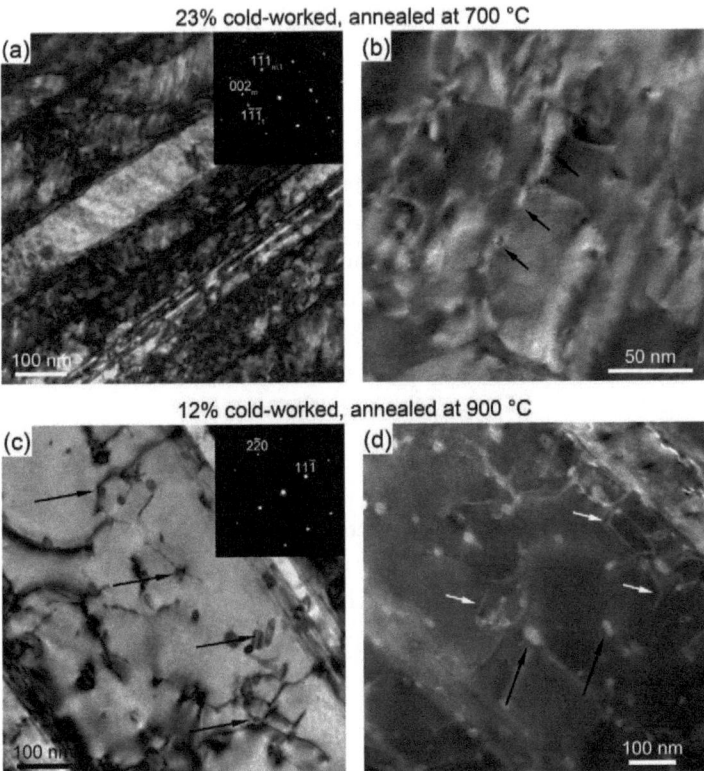

Figure 3.7: CTEM and STEM micrographs of TWIP-1Pd samples in the final heat treatment state, i.e. first recrystallized (56% cold-worked and annealed for 10 min at 1150 °C) and subsequently again cold-worked and annealed; see also Fig. 3.5. (a) Bright-field TEM image of a 23% cold-worked sample annealed for 30 min at 700 °C (TWIP-1Pd CW23-ht). The grain is oriented in the [1 1 0] zone-axis and shows a bundle of deformation twins running diagonally across the image. (b) Dark-field STEM micrograph of the same specimen. The short black arrows point to fringes of Moirée contrast which are all aligned parallel to each other and are located along a dislocation line exhibiting a bright contrast. (c) TEM bright-field image taken in the [1 1 2] zone-axis of a 12% cold-worked sample annealed for 30 min at 900 °C. Long black arrows point to the precipitates that formed upon the heat treatment. Traces of dislocations are seen as black lines that interact with the precipitates. (d) The HAADF-STEM image acquired from about the same area as in (c) shows that the precipitates are enriched with Pd (heaviest constituent of the alloy). The white arrows point to lines of bright contrast indicating enrichment of lattice defects with Pd.

69

3.3.3.2 Mechanical properties

The mechanical properties are summarized in Table 3.2, and a selection of representative stress-strain curves are depicted in Fig. 3.8. For comparison the mechanical properties of pure Fe (Armco quality) and 316L (stainless steel) in the annealed state are also shown. In the recrystallized state the TWIP alloys show similar mechanical properties, characterized by moderately high yield strengths (YS) of 350 MPa, pronounced work hardening generating high ultimate tensile strengths (UTS) of about 970 MPa and high uniform elongations (ϵ_U) of approximately 60%. Cold working generates an increase in YS and UTS at the expense of ductility, as indicated by the performance of the 12% and 23% cold-worked samples.

Upon annealing at 700 °C two processes occur: the formation of Pd-rich precipitates and recovery. The result is a significant decrease in the YS to 500 and 725 MPa for the 12% and 23% cold-worked samples, respectively. In the case of the samples deformed to 12%, the UTS decreases to 1020 MPa, whereas the uniform elongation increases to almost 50%. The UTS of the 23% cold-worked sample remains at the high level of 1250 MPa with a uniform elongation of 35%.

Table 3.2: Mechanical properties (from uniaxial tensile tests at room temperature) of the alloys investigated.

Alloy	YS (MPa)	UTS (MPa)	ϵ_U (%)	ϵ_f (%)
TWIP rexx	345 ± 10	980 ± 5	57 ± 3	62 ± 4
TWIP-1Pd rexx	360 ± 40	970 ± 35	60 ± 5	64 ± 3
TWIP-1Pd CW12	690 ± 35	1120 ± 25	36 ± 3	38 ± 3
TWIP-1Pd CW12-ht	505 ± 20	1020 ± 10	48 ± 5	53 ± 7
TWIP-1Pd CW23	1095 ± 35	1320 ± 15	24 ± 2	29 ± 2
TWIP-1Pd CW23-ht	725 ± 20	1255 ± 15	33 ± 1	38 ± 2
Fe (Armco)	230 ± 5	300 ± 5	19 ± 1	37 ± 1
316L	290 ± 15	630 ± 5	42 ± 1	62 ± 4

YS, yield strength; UTS, ultimate tensile strength; ϵ_U, uniform elongation; ϵ_f, elongation to fracture

3.4 Discussion

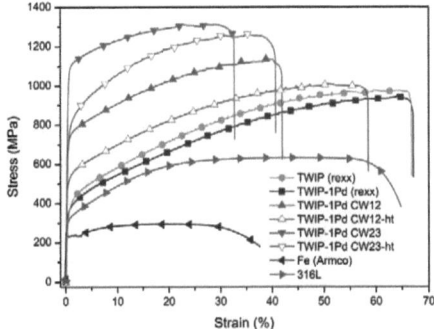

Figure 3.8: Stress-strain curves obtained from uniaxial tensile tests of the TWIP and TWIP-1Pd alloys in different heat treatment states. The TWIP and TWIP-1Pd alloys, especially in the recrystallized (rexx) and annealed (TWIP-1Pd CW12-ht and TWIP-1Pd CW23-ht) states, show a pronounced strain hardening due to the formation of deformation twins.

3.4 Discussion

3.4.1 Recrystallization behavior

3.4.1.1 Influence of annealing temperature at a constant degree of cold working

The main finding from the recrystallization experiments presented in Fig. 3.2a is that adding only 1 wt.% Pd (corresponding to approximately 0.5 at.%) significantly retards recrystallization of TWIP-1Pd compared to the alloy without Pd. The TWIP alloy showed complete recrystallization within 10 min at temperatures above 650 °C. This is in good agreement with studies on the recrystallization behavior of similar TWIP steels [17, 28-30]. In contrast, the recrystallization of TWIP-1Pd is clearly shifted to higher temperatures. It is known from aluminum alloys [31-33] and microalloyed steels [23, 25, 26] that precipitation and recrystallization interact and influence the resulting microstructure. Such interaction between the two processes can occur in the following ways. (i) The mobility of grain boundaries depends on the solute content of the matrix

[34, 35]. Foreign atoms segregate to the grain boundary and reduce its mobility. This is expected to retard recrystallization, but not completely inhibit it as its driving force is still active. (ii) Dispersed particles exhibit pinning forces on the recrystallization front, slowing the progress of recrystallization (Zener drag) [26, 27, 36]. Since Zener drag directly affects the residual driving force, the particles are in principle able to impede recrystallization completely. (iii) Strain-induced heterogeneous precipitation frequently occurs at dislocations in the matrix. A reduced dislocation density caused by recrystallization thus lowers the number of sites available for precipitation and hence retards its onset [26].

Additionally, it is also important to consider the interaction between recovery and concurrent precipitation [26, 37]. (i) Finely dispersed particles can effectively pin dislocation segments and hence retard recovery [26]. (ii) Alloying elements in solution (i.e. Pd in this study) decrease the mobility of dislocation by solute drag [26, 38]. The decreased dislocation mobility hinders the rearrangement and annihilation of dislocations to form low-angle boundaries. Because recrystallization usually initiates through local recovery, it is evident that retarded recovery also hinders the onset of recrystallization [37]. In the following, we discuss these processes for the TWIP-1Pd samples for the various annealing temperatures applied.

3.4.1.2 Model for the interplay between precipitation and recrystallization

The semi-quantitative diagram in Fig. 3.9 shows the proposed relation between recrystallization and precipitation, and highlights the mutual interaction between the two processes. It is based on full microstructure evaluation of all samples presented in this study and partly on studies of aluminum alloys [33, 37] and microalloyed steel [23]. The recrystallization start (R_s) and finish (R_f) lines are based on the data of the Pd-free TWIP alloy and on Ref. [28]. The symbol P_s indicates the onset of precipitation in the undeformed material. The shift of this curve to shorter times accounts for the accelerating effect of cold working on precipitation and is given by $P_{s,\text{def}}$.

3.4 Discussion

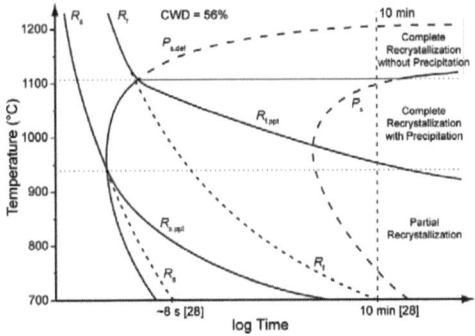

Figure 3.9: Schematic diagram illustrating the interaction between recrystallization and precipitation as a function of annealing time for TWIP-1Pd alloys. Recrystallization is characterized by the start (R_s) and finish (R_f) lines in the case of no interaction with precipitates and by the $R_{s,ppt}$- and $R_{f,ppt}$-lines in the case of interaction with precipitation. The latter process is characterized by the precipitation start line ($P_{s,def}$) for a cold-worked sample and the P_s-line in case of no prior deformation.

In accordance with the results presented in Fig. 3.2, three regimes can be distinguished: partial recrystallization and precipitation between 700 and 900 °C, complete recrystallization with precipitation between 950 and 1100 °C, and complete recrystallization without precipitation above 1100 °C.

In the temperature range between 700 and 900 °C precipitation ($P_{s,def}$-curve) takes place prior to the start of recrystallization. A large number of very fine Pd-rich precipitates forms in the deformed matrix and they retard recovery and the onset of recrystallization. Both the recrystallization start and finish are therefore delayed and are given by the $R_{s,ppt}$- and $R_{f,ppt}$-lines. The strong interplay between precipitation and recrystallization in this temperature range leads to only partial recrystallization within the observation period. In addition, concurrent recovery, which lowers the driving force for recrystallization, may eventually induce a complete halt of the latter process [25].

Complete recrystallization was only observed at temperatures above 950 °C. Between 950 and 1100 °C complete recrystallization, accompanied by the formation of Pd-rich precipitates was found within the observation period. The grain boundaries are decorated with relatively large Pd-rich precipitates in addition to small ones within the grains, as shown in Fig. 3.4b. Based on this observation, we conclude that recrystallization was initiated first, followed by precipitation. The precipitates are formed on the substructures of the deformed matrix and delay the recrystallization as described above. The recrystallization finish line is therefore shifted to longer annealing times and is here indicated by the $R_{f,ppt}$-line.

Above approximately 1100 °C and taking into account the accelerating effect of cold working, recrystallization is completed before precipitation initiates. Therefore no interaction is achieved. The precipitation in the defect-free recrystallized grains is indeed slow, as indicated by the P_s-curve, and would only be expected for long annealing times [23]. This, however, was not seen in this study within the observation period of 10 min. However, Pd is assumed to exhibit a solute drag effect on the grain boundary motion [35], leading to less pronounced grain coarsening compared to the Pd-free alloy.

The maximum solubility of Pd in the Fe–Mn–C matrix is assumed to be obtained between 1150 and 1200 °C. Above this temperature no precipitation can occur, even when assisted by the accelerating influence of cold working.

This model is partially based on investigations of microalloyed steels, where it was used to describe the interplay between recrystallization and the formation of niobium carbonitride precipitates [23]. The results shown in this study provide evidence that this model can be extended to TWIP steels alloyed with 1 wt.% Pd. The model presented in Fig. 3.9 successfully describes the mutual interplay between recrystallization and the formation of Pd-rich precipitates upon annealing of cold-worked TWIP-1Pd steel. It may be valid not only for the particular case of Pd alloying, but can also be adapted to describe the behavior of other TWIP steels, e.g. those involving the widely-used precipitate-forming alloying elements Nb, Ti or V [17]. Adapting the model presented here to the cases of other alloying systems may lead to a better understanding of the interplay between recrystallization and precipitation. This knowledge facilitates a more efficient design of thermo-mechanical treatments to optimize the alloys' performance.

3.4.1.3 Influence of the degree of cold working

The results presented in Figs. 3.2, 3.3 and 3.4 and the model developed from this data is for a given degree of cold working: 56%. The amount of deformation prior to annealing strongly influences the kinetics of recrystallization [37]. A lower CWD leads to a deceleration of recrystallization. In the model shown in Fig. 3.9, this corresponds to a shift of the recrystallization start- and finish-lines to longer times. Similarly, the onset of precipitation occurs earlier with an increasing CWD, as this corresponds to a higher number of possible nucleation sites [26, 27]. However, we assume that the influence of cold working on the recrystallization kinetics is more pronounced than on the precipitation reaction. This is supported by the data presented in Fig. 3.5. For a given annealing temperature, e.g. 700 °C, a minimum CWD of approximately 30% is required to initiate recrystallization. This means that for samples cold-worked below this threshold, no recrystallization takes place upon annealing at this temperature. Considering the model shown in Fig. 3.9, it is assumed that precipitation still occurs and is accompanied by a substantial amount of recovery. This offers a possibility to specifically adjust microstructure and hence both mechanical and electrochemical properties. In fact, the final thermo-mechanical optimization treatment makes use of these findings: the degrees of cold working of 12% and 23% were chosen so that no recrystallization occurs during the final annealing treatment. Nevertheless, the cold working and hence the increased dislocation density makes possible the formation of Pd-rich precipitates. The results presented in Fig. 3.5 extend the model shown in Fig. 3.9 by including the influence of the degree of cold working. This renders the model even more useful for efficiently designing thermo-mechanical treatments, which will be discussed in more detail in the next section.

3.4.2 Thermo-mechanical optimization

3.4.2.1 Precipitate evolution

By comparing the microstructure in Fig. 3.7b (sample annealed at 700 °C) to those in Figs. 3.7c and 3.7d (specimens annealed at 900 °C), it is apparent that a higher annealing temperature results in considerably larger precipitates. The DF-STEM image in Fig. 3.7b shows approximately 4 nm large islands of Moirée contrast. The actual size of the precipitates, however, is smaller than the size of those islands. The island fringes are oriented in a small number of directions which indicates that the precipitates might be at least semi-coherent to the matrix. Figs. 3.7c and d show the microstructure of a sample annealed at 900 °C. The precipitates are elongated and significantly larger, i.e. about 20 to 50 nm. They are still oriented in a small number of directions within the same grain, providing evidence of a definite crystallographic correspondence with the matrix.

The precipitates are arranged along dislocation lines. We thus assume that pipe diffusion of Pd atoms along the dislocation cores assists the precipitation process. This is supported by the observation of Pd-enriched dislocation cores, which are clearly visible in the atomic-number-sensitive STEM-HAADF imaging mode (Fig. 3.7d, short white arrows). Pipe-diffusion-assisted precipitation is well-known for maraging steels [21] and microalloyed steels [23, 24], and has also recently been reported for martensitic Fe–Mn–Pd alloys [8].

As pointed out above, the precipitates interact with the dislocations present in the material (Figs. 3.7b and c). Particularly in the case of the sample annealed at 700 °C, it is expected that such a precipitate-dislocation arrangement will be favorable because it reduces the coherency stresses at the particle-matrix interface. Even the larger precipitates found after annealing at 900 °C are still effective at pinning the dislocations, as seen in Fig. 3.7c.

Only a few attempts have been made to use precipitation strengthening in TWIP steels [17, 39, 40]. The alloying elements Nb, Ti and V, known from microalloyed steels to form precipitates in the austenite phase, have been used to investigate the possible precipitation strengthening in Fe-22Mn-0.6C [39]. The strengthening effect due to the formation of fine Ti-containing precipitates (2 nm in size) was about 75 MPa and was explained via the Ashby–Orowan approach [39]. In this study the small Pd-rich precipitates, formed in particular after annealing the TWIP-1Pd alloy at 700 °C, may contribute with a similar amount to the strength of the alloys, as discussed in the next section.

3.4.2.2 Mechanical performance

The mechanical performance of TWIP and TWIP-1Pd in the recrystallized state is comparable to the behavior reported by many authors for similar alloys, e.g. [17, 20, 39]. Cold working generates an increase in strength at the expense of reduced ductility. Upon annealing at 700 °C two processes occur: the formation of Pd-rich precipitates, and recovery, which can be identified by the significant drop in the YS. The mechanical performance of the 12% cold-worked and annealed sample is in principal similar to that of the recrystallized state. However, the YS is approximately 150 MPa higher and the work hardening deviates by approximately 30% strain from the behavior of the recrystallized samples. The increase in the YS can be attributed to the hardening effect of the Pd-rich precipitates. The deviation of the strain hardening response has been observed previously and attributed to a reduced rate of stacking fault and twin boundary formation [39]. Even though the underlying physical mechanism which controls the interaction between precipitates and twinning is not known, this result indicates an earlier saturation of twin formation in the precipitate-containing alloy than in the alloy without precipitates [39]. As the main feature leading to the high strain hardening rate of these steels is the formation of twins upon deformation [17], it may be speculated that the precipitates impede the motion of the Shockley partial dislocations needed for the formation of the deformation twins. This assumption is supported by the observation of the pronounced interaction between the Pd-rich precipitates and dislocations, as shown in Fig. 3.7.

In terms of application as biodegradable stent material, a high-strength material featuring a pronounced strain hardening is desirable as this facilitates filigree implant designs and supports safe stent placement (i.e. homogeneous dilata-

3 Microstructure and mechanical performance

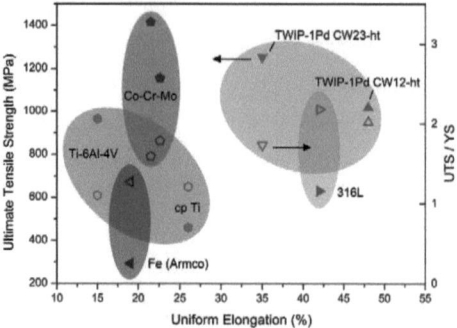

Figure 3.10: Diagram comparing the mechanical properties of the TWIP(-1Pd) alloys to alloys currently in use for permanent medical implants. The alloy's performance is expressed in terms of both the UTS (solid symbols) and the ratio of UTS/YS (as a measure of the strain hardening capability, open symbols) over uniform elongation. The TWIP-1Pd alloys investigated here combine the high strength of Co–Cr–Mo alloys with the high ductility of stainless steel 316L.

tion). In this respect, the mechanical performance of the 23% cold-worked and annealed samples is especially interesting. They show a moderately high YS, considerable strain hardening generating a high UTS, and a uniform elongation of 35%. Fig. 3.10 shows a comparative illustration of the UTS and the ratio UTS / YS as a function of uniform elongation for TWIP(-1Pd), pure Fe and three materials typically used nowadays for permanent implants: stainless steel 316L, Ti / Ti-6V-4Al [41] and Co–Cr–Mo alloys [42]. An ideal stent-material should feature a high UTS as well as a high ratio of UTS/YS at high uniform ductility. (The ratio UTS/YS is taken as a measure of the strain hardening capability of a given alloy.) Pure Fe, Ti / Ti-6V-4Al and Co–Cr–Mo alloys mainly suffer from rather low uniform elongation even though Ti-6V-4Al and Co–Cr–Mo materials show high strength values. Stainless steel 316L features extended uniform elongation, but the UTS is only moderate. In contrast, the heat-treated TWIP-1Pd steels offer a considerably more attractive combination of mechanical properties, i.e. an UTS of approximately 1000 MPa (comparable to Ti-6V-4Al) or more, combined with a large uniform elongation and a high strain hardening capability. Given that it was possible to successfully produce and implant stents made of pure iron [3, 4], the mechanical performance of these TWIP-1Pd alloys is very promising for their use as degradable stents, providing maximum freedom in their design. The use of this material, however, is not only limited to stents, but may be extended to osteosynthesis applications, which also require high-strength materials with appropriate ductility.

3.5 Conclusions

TWIP steels offer advantageous mechanical properties which make them very attractive as biodegradable implant materials. The particular composition of the alloys investigated (Fe-21Mn-0.7C(-1Pd)) was chosen according to the considerations formulated in the design strategy, aiming at both favorable mechanical performance and an increased corrosion rate. It was shown in this study that the addition of 1 wt.% Pd not only influences the corrosion properties but also has a strong influence on the microstructure and the mechanical properties. During annealing of cold-worked samples, both recrystallization and the formation of Pd-rich precipitates take place, and the two processes strongly interact with each other. The fine precipitates formed in the deformed matrix during annealing provide a pinning force which restricts the grain boundary motion (Zener drag). It was also shown that the dislocation cores are enriched with Pd, due to preferential diffusion along dislocation cores (pipe diffusion). This generates decreased dislocation mobility due to a solute drag effect, significantly retarding recovery and recrystallization in comparison with the Pd-free alloys.

Based on the experimental observations, a semi-quantitative model which allows us to understand the interplay between recrystallization and precipitation was established. This model explains the different microstructures observed after annealing at different temperatures. It is expected that the model presented in this study can be extended to predict the behavior of other TWIP steels alloyed with precipitate-forming elements. The insights gained may help to improve thermo-mechanical treatments for optimizing microstructure and mechanical performance of TWIP steels.

Taking into account the impeding effect of the precipitates on recrystallization, a thermo-mechanical treatment was explored in order to improve the mechanical performance with respect to potential application as degradable implant material. Here the alloys were slightly cold-worked and subsequently annealed at moderate temperature (i.e. 700 °C). The very fine Pd-rich precipitates that formed in the deformed matrix efficiently suppressed the onset of recrystallization. However, the alloys showed a considerable amount of recovery, which generated a microstructure that is in good agreement with the requirements previously formulated in the design strategy. The mechanical performance is considered interesting for temporary implant applications due to the favorable combination of high strength and ductility with a pronounced strain hardening.

Acknowledgements

The authors thank A. Wyss for support with the tensile tests; A.S. Sologubenko, E. Müller and F. Gramm for support regarding TEM and for stimulating discussions; and Ch. Wegmann for help with the sample preparation. The support of the ETH Zurich Electron Microscopy Center is gratefully acknowledged. The authors appreciate the financial support received within the framework of the project "Biocompatible Materials and Applications" initiated by the Austrian Institute of Technology GmbH (AIT), and support from the Staub/Kaiser Foundation, Switzerland.

References

[1] Moravej M, Mantovani D. Biodegradable Metals for Cardiovascular Stent Application: Interests and New Opportunities. Int J Mol Sci 2011;12:4250-70.

[2] Hermawan H, Dubé D, Mantovani D. Developments in metallic biodegradable stents. Acta Biomater 2010;6:1693-7.

References

[3] Peuster M, Hesse C, Schloo T, Fink C, Beerbaum P, von Schnakenburg C. Long-term biocompatibility of a corrodible peripheral iron stent in the porcine descending aorta. Biomaterials 2006;27:4955-62.

[4] Peuster M, Wohlsein P, Brugmann M, Ehlerding M, Seidler K, Fink C, et al. A novel approach to temporary stenting: degradable cardiovascular stents produced from corrodible metal - results 6-18 months after implantation into New Zealand white rabbits. Heart 2001;86:563-9.

[5] Waksman R, Pakala R, Baffour R, Seabron R, Hellinga D, Tio FO. Short-term effects of biocorrodible iron stents in porcine coronary arteries. J Interv Cardiol 2008;21:15-20.

[6] Hermawan H, Alamdari H, Mantovani D, Dube D. Iron-manganese: new class of metallic degradable biomaterials prepared by powder metallurgy. Powder Metall 2008;51:38-45.

[7] Schinhammer M, Hänzi AC, Löffler JF, Uggowitzer PJ. Design strategy for biodegradable Fe-based alloys for medical applications. Acta Biomater 2010;6:1705-13.

[8] Moszner F, Sologubenko AS, Schinhammer M, Lerchbacher C, Hänzi AC, Leitner H, et al. Precipitation hardening of biodegradable Fe–Mn–Pd alloys. Acta Mater 2011;59:981-91.

[9] Schumann H. Die martensitischen Umwandlungen in kohlenstoffarmen Manganstählen. Arch Eisenhüttenw 1967;38:647-56.

[10] Rémy L, Pineau A. Twinning and strain-induced F.C.C. → H.C.P. transformation in the Fe–Mn–Cr–C system. Mater Sci Eng 1977;28:99-107.

[11] Tomota Y, Strum M, Morris J. Microstructural dependence of Fe-high Mn tensile behavior. Metall Mater Trans A 1986;17:537-47.

[12] Shiang LT, Wayman CM. Maraging behavior in an Fe-19.5Ni-5Mn alloy I: Precipitation characteristics. Metallography 1988;21:399-423.

[13] Frommeyer G, Brux U, Neumann P. Supra-ductile and high-strength manganese-TRIP/TWIP steels for high energy absorption purposes. ISIJ Int 2003;43:438-46.

[14] Allain S, Chateau JP, Bouaziz O, Migot S, Guelton N. Correlations between the calculated stacking fault energy and the plasticity mechanisms in Fe–Mn–C alloys. Mater Sci Eng 2004;387-389:158-62.

[15] Scott C, Allain S, Faral M, Guelton N. The development of a new Fe–Mn–C austenitic steel for automotive applications. Rev Métall - CIT 2006;103:293-302.

[16] Ueji R, Tsuchida N, Terada D, Tsuji N, Tanaka Y, Takemura A, et al. Tensile properties and twinning behavior of high manganese austenitic steel with fine-grained structure. Scripta Mater 2008;59:963-6.

[17] Bouaziz O, Allain S, Scott CP, Cugy P, Barbier D. High manganese austenitic twinning induced plasticity steels: A review of the microstructure properties relationships. Curr Opin Solid State Mater Sci 2011;15:141-68.

[18] Hamada AS, Karjalainen LP. Hot ductility behaviour of high-Mn TWIP steels. Mater Sci Eng A 2011;528:1819-27.

[19] Jin J-E, Lee Y-K. Strain hardening behavior of a Fe-18Mn-0.6C-1.5Al TWIP steel. Mater Sci Eng A 2009;527:157-61.

[20] Santos DB, Saleh AA, Gazder AA, Carman A, Duarte DM, Ribeiro ÉAS, et al. Effect of annealing on the microstructure and mechanical properties of cold rolled Fe-24Mn-3Al-2Si-1Ni-0.06C TWIP steel. Mater Sci Eng A 2011;528:3545-55.

[21] Vasudevan V, Kim S, Wayman C. Precipitation reactions and strengthening behavior in 18 Wt Pct nickel maraging steels. Metall Mater Trans A 1990;21:2655-68.

[22] Zurob HS, Brechet Y, Purdy G. A model for the competition of precipitation and recrystallization in deformed austenite. Acta Mater 2001;49:4183-90.

[23] Hansen S, Sande JB, Cohen M. Niobium carbonitride precipitation and austenite recrystallization in hot-rolled microalloyed steels. Metall Mater Trans A 1980;11:387-402.

[24] Dutta B, Palmiere EJ, Sellars CM. Modelling the kinetics of strain induced precipitation in Nb microalloyed steels. Acta Mater 2001;49:785-94.

[25] Vervynckt S, Verbeken K, Thibaux P, Houbaert Y. Recrystallization-precipitation interaction during austenite hot deformation of a Nb microalloyed steel. Mater Sci Eng A 2011;528:5519-28.

[26] Zurob HS, Hutchinson CR, Brechet Y, Purdy G. Modeling recrystallization of microalloyed austenite: effect of coupling recovery, precipitation and recrystallization. Acta Mater 2002;50:3077-94.

[27] Dutta B, Sellars CM. Effect of composition and process variables on Nb(C, N) precipitation in niobium microalloyed austenite. Mater Sci Technol 1987;3:197-206.

[28] Lü Y, Molodov DA, Gottstein G. Recrystallization kinetics and microstructure evolution during annealing of a cold-rolled Fe–Mn–C alloy. Acta Mater 2011;59:3229-43.

[29] Bracke L, Verbeken K, Kestens L, Penning J. Microstructure and texture evolution during cold rolling and annealing of a high Mn TWIP steel. Acta Mater 2009;57:1512-24.

[30] Kang S, Jung Y-S, Jun J-H, Lee Y-K. Effects of recrystallization annealing temperature on carbide precipitation, microstructure, and mechanical properties in Fe-18Mn-0.6C-1.5Al TWIP steel. Mater Sci Eng A 2010;527:745-51.

[31] Verdier M, Brechet Y, Guyot P. Recovery of AlMg alloys: flow stress and strain-hardening properties. Acta Mater 1998;47:127-34.

[32] Deschamps A, Brechet Y. Influence of predeformation and ageing of an Al–Zn–Mg alloy - II. Modeling of precipitation kinetics and yield stress. Acta Mater 1998;47:293-305.

[33] Jones MJ, Humphreys FJ. Interaction of recrystallization and precipitation: The effect of Al_3Sc on the recrystallization behaviour of deformed aluminium. Acta Mater 2003;51:2149-59.

[34] Rutter JW, Aust KT. Kinetics of Grain Boundary Migration in high-purity Lead containing very small Additions of Silver and of Gold. Trans Am Inst Min Metall Eng 1960;218:682-8.

[35] Cahn JW. The impurity-drag effect in grain boundary motion. Acta Metall 1962;10:789-98.

[36] Manohar PA, Ferry M, Chandra T. Five decades of the Zener equation. ISIJ Int 1998;38:913-24.

[37] Humphreys FJ, Hatherly M. Recrystallization and related annealing phenomena: Elsevier: Amsterdam; 2004.

[38] Arieta FG, Sellars CM. Activation volume and activation energy for deformation of Nb HSLA steels. Scripta Metall Mater 1994;30:707-12.

[39] Scott C, Remy B, Collett J-L, Cael A, Bao C, Danoix F, et al. Precipitation strengthening in high manganese austenitic TWIP steels. Int J Mater Res 2011;102:538-49.

[40] Yazawa Y, Furuhara T, Maki T. Effect of matrix recrystallization on morphology, crystallography and coarsening behavior of vanadium carbide in austenite. Acta Mater 2004;52:3727-36.

[41] Rack HJ, Qazi JI. Titanium alloys for biomedical applications. Mater Sci Eng C 2006;26:1269-77.

3 Microstructure and mechanical performance

[42] Chiba A, Kazushige K, Takeda H, Nomura N. Mechanical Properties of Forged Low Ni and C-Containing Co–Cr–Mo Biomedical Implant Alloy. Mater Sci Forum 2005;475 - 479:2317 - 22.

4 Degradation properties I - Immersion testing

Assessing the degradation performance of materials being considered for temporary medical implants is very important for their application. However, the correlation between laboratory tests such as immersion testing and electrochemical methods and in vivo results is still unsatisfactory. In particular, adding a pH buffer generally increases degradation, which results in an overestimation of in vivo rates. Chapter 4 presents results from immersion tests in a simulated body fluid whose pH value is solely controlled using gaseous CO_2. It was found that this system more accurately reflects the in vivo system for well-vascularized tissues.

On the immersion testing of degradable implant materials in simulated body fluid[1]

Predictions of in vivo degradation behavior derived from laboratory experiments are of great importance in the development of biodegradable materials. A key issue is the simulation of the physiological conditions found in living organisms. Generally, the testing solution and in particular the pH buffer have a significant influence on the outcome of degradation experiments. The addition of pH buffers (such as Tris or Hepes) to simulated body fluid (SBF) has been found to greatly accelerate degradation, in turn generating an overestimation of the actual degradation rate.

This study presents results obtained from immersion tests in SBF buffered with gaseous CO_2. The control of the pH value by means of CO_2 has the advantage of better reflecting physiological conditions, and allows the carrying out of in vitro experiments which are closer to the in vivo situation. In addition, because of its modular design this approach can be easily adapted and combined with other experimental techniques. For the magnesium alloy WZ21, the degradation rate was found to correlate well with results from in vivo studies, illustrating the potential of this approach.

4.1 Introduction

The characterization of degradation properties is of paramount importance in the development and evaluation of materials for temporary medical applications, e.g. osteosynthesis implants or cardiovascular stents. Magnesium and iron alloys are currently the focus of ongoing research [1-9], and one of the major challenges to be dealt with is the establishment of a reliable quantitative correlation between in vitro and in vivo degradation results [5, 10, 11]. Only a laboratory method which yields a prediction on in vivo performance can make

---

[1]M. Schinhammer, J. Hofstetter, Ch. Wegmann, F. Moszner, J.F. Löffler, P.J. Uggowitzer; Advanced Engineering Materials doi: 10.1002/adem.201200218

possible efficient material evaluation. Different approaches to drawing up such a method have been presented in the literature. Often, simulated physiological media such as Hank's solution or simulated body fluid (SBF) have been employed [12-19]. Recent developments aim at simulating the in vivo situation closely, thereby mimicking as many living organism characteristics as possible. In consequence the complexity of test setups has generally increased, e.g. in the context of simulating arterial blood flow [18], adding proteins to SBF [20], applying cell culture conditions [20] or controlling the oxygen [21] or carbon dioxide concentrations [13]. Despite these efforts, a reliable quantitative correlation between in vitro and in vivo results is not yet available.

Among other challenges, maintaining a constant pH value in the testing environment is currently limiting the significance of laboratory experiments. Upon Mg degradation, HO^- ions are generated which cause the pH value to increase and rapidly exceed the physiological range (of blood plasma) of 7.40 ± 0.05 [22]. A large number of studies reports pH values as high as 10.5 (at which Mg forms a stable oxide layer), thus not sufficiently reflecting the situation in living organisms [19, 23-28]. In order to achieve more constant and stable conditions, a pH buffer (e.g. Tris or Hepes) is often added. However, this measure only delays, but is not able to prevent, the pH increase (due to a limited buffering capacity), and because such a buffer is not present in the living organism it is desirable to completely avoid using it. The most important pH buffer system in the body is the carbonate buffer [22, 29]. The organism produces a significant acid load each day in the form of volatile CO_2. The CO_2 reacts with water to form carbonic acid (H_2CO_3) which is again in equilibrium with HCO_3^-:

$$CO_2 + H_2O \rightleftarrows H_2CO_3 \rightleftarrows H^+ + HCO_3^- \qquad (4.1)$$

It is convenient and highly effective to adapt this mechanism to counteract the pH increase due to Mg degradation. In the present setup we feed gaseous CO_2 into the testing solution and actively regulate the pH value. This approach has two important advantages: (i) it allows control of the pH value in a narrow range, and (ii) it makes use of a pH buffer obsolete. Active control of the pH value is especially beneficial for the investigation of Mg degradation, but is not limited to it. Generally, the pH value increases gradually in any carbonate-containing solution because gaseous CO_2 leaves the solution. Therefore the approach is also advantageous in the case of degradable Fe.

In this study we present results from immersion tests with active pH control using CO_2. We compare these results with those obtained using Tris- and Hepes-buffered solutions and recent data from in vivo studies, and discuss the influence of the buffering system on the reliability of immersion tests.

4.2 Materials and Methods

4.2.1 Materials

In this study two alloys representative of biodegradable metals were investigated: pure iron (Armco quality) and the degradable magnesium alloy WZ21 (Mg-2Y-1Zn-0.25Ca-0.15Mn, in wt.%). Pure Fe was chosen to represent the class of biodegradable Fe-based alloys. The processing route, microstructure and mechanical performance of the Mg alloy, which was used in the as-extruded state, are described in Ref. [30]. This alloy was developed for stent applications and shows a good combination of strength and ductility. It was particularly selected because both in vitro and in vivo results are available [6, 12, 31].

Magnesium corrosion can be readily evaluated via the hydrogen evolution method [12, 32]: As every mole of Mg produces one mole of hydrogen, it is possible to derive the degradation rate from the amount of hydrogen evolved. The hydrogen evolution data from Hänzi et al. [12] were taken as the reference for performance in simulated physiological solutions with different buffering capacities, i.e. SBF buffered using Tris (tris(hydroxymethyl)aminomethane, denoted as SBF(Tris)), phosphate buffered saline (PBS) and minimum essential medium (α-MEM). All testing solutions were kept at 37 ± 2 °C [12]. SBF(Tris) shows the highest buffering capacity among these three solutions, owing to the Tris buffer (concentration of 50 mmol·l^{-1}). The lowest buffering capacity is found in α-MEM [12]. In this study we also present results from immersion tests using SBF buffered with Hepes (4-(2-hydroxyethyl)-1-piperazineethanesulfonic acid, Carl Roth GmbH, SBF(Hepes)) in order to gain further insights on the influence of the pH buffer. The tests were performed as previously described [12]; the ion concentrations of the solutions are summarized in Table 4.1. The data gained from in vivo microfocus computed tomography (μCT) analysis was used to determine the degradation performance of WZ21 in living organisms [6].

Table 4.1: Ion concentrations and pH values of the simulated physiological media discussed in this study (concentrations are indicated in mmol·l^{-1}).

Ions	SBF	PBS	α-MEM
Na$^+$	142.0	153.0	118.3
K$^+$	5.0	5.0	5.3
Mg^{2+}	1.0	—	0.8
Ca^{2+}	2.5	—	1.8
Cl$^-$	109.0	140.0	126.2
HCO$_3^-$	27.0	—	26.2
HPO$_4^{2-}$	1.0	8.0	—
H$_2$PO$_4^-$	—	2.0	1.0
SO$_4^{2-}$	1.0	—	0.8
pH	7.3–7.4	7.35	7.5

SBF: simulated body fluid,
PBS: phosphate buffered saline,
α-MEM: minimum essential medium

4.2.2 Immersion testing setup

A schematic representation of the test setup deployed for the Fe-based alloys is shown in Fig. 4.1a. The glass container which holds the samples and the testing solution (kept at 37 ± 1 °C) is placed on a magnetic stirrer, used to agitate the solution and to control its temperature. This ensures constant conditions throughout the testing medium. The samples are placed on a sample holder made from

4 Degradation properties I - Immersion testing

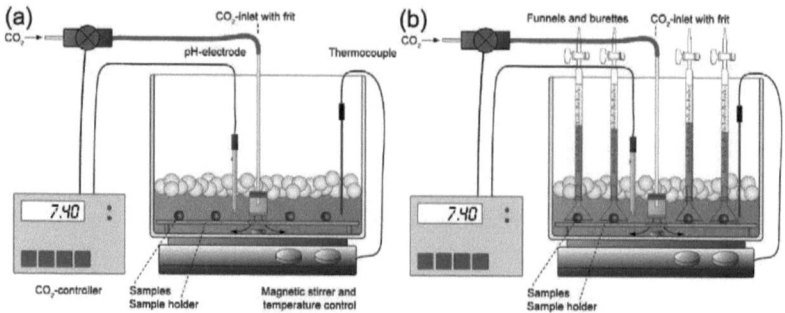

Figure 4.1: Schematic illustration showing the testing setup used in this study. (a) The setup for Fe samples consists of a glass container which contains the testing solutions and the samples. A magnetic stirrer is used to control the temperature and to agitate the solution. The pH value is monitored by a programmable threshold switch and opens a solenoid valve once the pH value exceeds a given upper threshold value. Gaseous CO_2 is then fed via a frit into the testing solution until the lower threshold is reached. (b) The setup for Mg samples was additionally equipped with funnels and burettes to collect the hydrogen which evolves during degradation.

polycarbonate. The testing solution is covered with floating balls to reduce water evaporation and energy loss. A programmable threshold switch monitors the pH values of the testing medium by means of a commercially available pH electrode. Once the pH value of the testing medium exceeds a given threshold (i.e. 7.45), the switch opens a solenoid valve to feed gaseous CO_2 into the testing medium. The valve is closed once the lower threshold of 7.40 has been reached. In order to achieve an optimal dissolution of the CO_2, the inlet is equipped with a frit, centered over the magnetic stirrer. The gaseous CO_2 fed into the testing solution forms into small bubbles, which facilitates dissolving of the gas.

The SBF used in this study simulates the ionic composition and pH value of human blood plasma. Its composition is the same as that used in previous studies [12, 33], except that no pH buffer (Tris or Hepes) is used. The pH value was adjusted only via the CO_2 regulation, and neither HCl nor NaOH was used. In the following, this SBF is denoted as SBF(CO_2).

In order to accurately determine the mass loss of the iron samples the degradation products were removed, as follows. After immersion, the loose degradation products were dislodged using a brush. The samples were then ultrasonically cleaned in ethanol and subsequently immersed for 20 min in a solution composed of 11.1 g sodium dihydrogen citrate (ABCR GmbH), 1.7 g citric acid (Merck) and 0.1 g tryptamine (as corrosion inhibitor [34], ABCR GmbH) in 250 ml deionized water. They were mechanically cleaned once more using a brush and finally ultrasonically cleaned in ethanol and dried in hot air.

The degradation rate of the Mg samples can be determined conveniently using the hydrogen evolution method, which has been widely used in literature [5, 12, 32]. The testing setup was therefore extended by funnels and burettes. These are placed over each sample to collect the evolving hydrogen, as indicated in Fig. 4.1b. The amount of hydrogen collected in the burette over time reveals the degradation rate of the Mg sample. Special care was taken to ensure that no CO_2 directly enters the funnels and burettes as this would bias the results. Therefore, the sample holder is made from a solid sheet of polycarbonate in the case of the Mg testing. The flow generated by the magnetic stirrer causes the CO_2 bubbles to move radial outwards between the sample holder and the glass container. Hence, the CO_2 bubbles have no possibility to directly go into the funnels and burettes. However, two more influences have to be considered: (i) CO_2 might be liberated at the liquid-air interface in the burettes, and (ii) a varying air pressure causes changes in the liquid level in the burettes. Both can be circumvented by comparing the actual hydrogen level to a blind from a burette without sample.

As pointed out by Kirkland et al. [5], it is important to use a sufficiently large solution-volume-to-sample-surface ratio (> 50 ml·cm^{-2}), which was 67 ml·cm^{-2} for the Mg samples and 160 ml·cm^{-2} for the Fe samples.

4.3 Results

4.3.1 Immersion testing of pure iron

The results of the immersion testing of pure Fe are shown in Fig. 4.2. The mass loss of the samples increased steadily over the period of 4 weeks. However, when taking into account the mass loss rate, it is apparent that the degradation rate slightly decreased with prolonged immersion time.

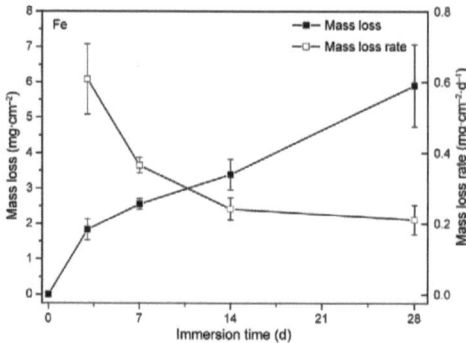

Figure 4.2: Mass loss and mass loss rate as function of immersion time in SBF(CO_2) for pure iron.

4.3.2 Immersion testing of WZ21

Hydrogen evolution versus immersion time is presented in Fig. 4.3a and the pH change of the corresponding media is shown in Fig. 4.3b. The hydrogen evolution of WZ21 in SBF(CO_2) is a linear function of the immersion time. In comparison to previous results [12], the hydrogen evolution rate lies between the performance in α-MEM and PBS. It is therefore markedly lower compared to the immersion in SBF(Tris). Considering the corresponding pH values of the SBF(CO_2) this finding is surprising: during the tests, the pH value never deviated from the range between 7.35 and 7.45 and was therefore constant and lower compared to those of the other three testing solutions. Due to their limited buffering capacity, the pH values of the latter three quickly increased during immersion testing. In

4.3 Results

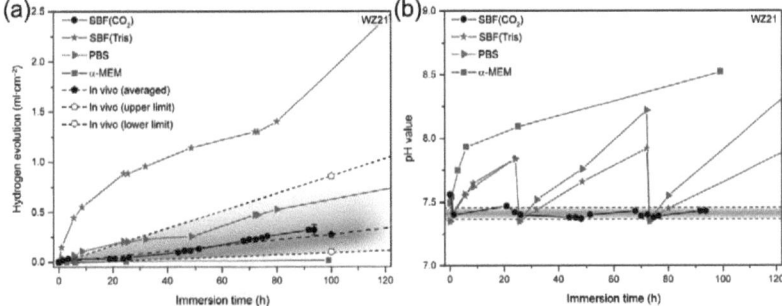

Figure 4.3: (a) Hydrogen evolution over immersion time of WZ21 immersed in different physiological media (SBF(CO_2), SBF(Tris), PBS and α-MEM) at 37 °C. The range of vivo degradation rates (average, lower and upper rate) based on μCT data is also indicated. The average in vivo degradation rate (given by the solid symbol) agrees well with the in vitro degradation performance in SBF(CO_2). The lower and upper in vivo degradation rates (open symbols) specify the range found in the animal study. (b) Corresponding evolution of the pH values of the testing solutions as function of immersion time. The pH value of SBF(CO_2) was rather constant and always in the range between 7.35 and 7.45.

order to maintain conditions as constant as possible, in [12] the testing solutions were regularly changed, which explains the drops in the pH values in Fig. 4.3b after 24 h and 72 h (for SBF(Tris) and PBS).

The μCT data presented in the in vivo study (Fig. 1a in [6]) shows the loss in implant volume over time. These values can easily be converted in equivalent volumes of evolved hydrogen that allow a comparison with the hydrogen evolution data from [12] and the present study. A set of three degradation rates (average, lower and upper limiting rate) was determined; these are plotted as linear functions in Fig. 4.3a. The reason for specifying a range of degradation rates instead of a single value is discussed in Section 4.4.3. The degradation performance of WZ21 in SBF(CO_2) and in PBS lies within the range given by the in vivo data. It is worth noting that the hydrogen evolution rate of WZ21 in SBF(CO_2) is quite close to the average in vivo degradation rate.

4 Degradation properties I - Immersion testing

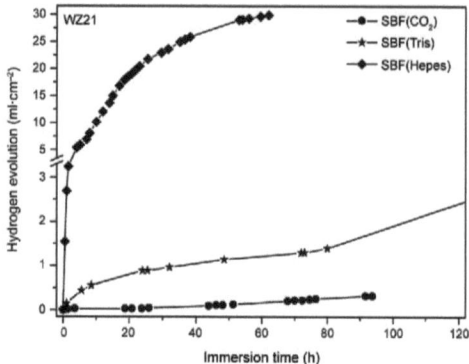

Figure 4.4: Hydrogen evolution as a function of immersion time of WZ21 immersed in differently buffered SBF (SBF(CO_2), SBF(Tris) and SBF(Hepes)).

The influence of the pH buffer is further elucidated by the results presented in Fig. 4.4. It is apparent that the choice and concentration of the pH buffer has a tremendous influence on the degradation rate. We used a simple linear approximation to determine the degradation rate in the different testing solutions. The degradation of WZ21 in SBF(Tris) is about five times faster than that of SBF(CO_2). When replacing Tris (at a concentration of 50 mmol·l^{-1}) with Hepes (100 mmol·l^{-1}) as a pH buffer, the degradation rate increases by a factor of approximately 60. For SBF(Hepes) the degradation rate was determined in the initial linear part of the curve. The subsequent decrease in the degradation rate can be attributed to the formation of a thick layer of degradation products, which slow down the process [12, 15].

4.4 Discussion

The ability to perform immersion tests at constant pH values without using a buffer is the main advantage of the testing procedure used in this study. The absence of a pH buffer has implications for the results obtained by immersion tests and their significance. These aspects are discussed in the following sections.

4.4 Discussion

4.4.1 The use of CO_2 as a pH buffer

Using the carbonate system to maintain a constant pH value in the testing solution corresponds to the situation in the human body: the main buffer system in blood plasma is the equilibrium between bicarbonate and carbonic acid [29]. Because there is usually only a low concentration of carbonic acid in the plasma, it is valid to consider the bicarbonate / CO_2 system instead [29]. Further buffer systems in plasma are plasma proteins and hemoglobin. Their importance, however, is limited compared to that of the carbonate system. The physiological disturbance in acid-base balance corresponding to Mg degradation would be a metabolic alkalosis, i.e. the removal of H^+ in the plasma. The change in pH is partially compensated by non-carbonate buffers and partially by alveolar hypoventilation, i.e. a reduced respiratory removal of CO_2 in the lungs [29]. In terms of eq. 4.1, the removal of H^+ corresponds to a shift of the equilibrium to the right-hand side of the equation and can be compensated by an increase in the CO_2 partial pressure. This is precisely the approach pursued in the immersion testing setup used in this study. The body additionally compensates the increase of the HCO_3^- concentration by renal excretion. This is not yet possible in the current setup and results in a slightly increased HCO_3^- concentration in the SBF during the in vitro tests. This in turn may increase the tendency for carbonate precipitation [2, 33, 35, 36]. However, for WZ21 in SBF(CO_2), the hydrogen evolved is approximately 0.32 ml·cm^{-2} in 94 h. Taking into account the amount of CO_2 necessary to compensate the pH increase and the experimental conditions (sample size and volume of solution), the resulting increase in the HCO_3^- concentration is moderate and only 0.4 mmol·l^{-1}. Thereby the HCO_3^- concentration in the SBF increases from 27.0 to 27.4 mmol·l^{-1}. This is not considered to influence the degradation rate of the samples by additional precipitation.

The use of CO_2 to control the pH value of the testing solution is a relatively simple, yet powerful method for achieving physiological conditions in degradation tests. Compared to other approaches, this procedure completely avoids using a pH buffer, while the pH value is maintained in a narrow interval around 7.40. One of the limitations of long-term experiments (such as mass-loss determination or hydrogen evolution measurements) is indeed the change in pH value [5]. This makes it necessary to either repeatedly replace the testing solution (to maintain a "constant" pH value) or otherwise increasing pH values lower the significance of the results because the testing conditions are far from physiological.

At first glance, this approach appears straightforward. However, its successful implementation requires consideration of the experimental challenges. It is important to ensure a good dissolution of the CO_2 in the testing solution. This is ensured by the flow conditions in the testing container: during the inflow of CO_2, the bubbles which emerge from the frit are dragged downwards to increase their contact time with the testing solution.

The system inherently possesses a relatively high inertia, i.e. after opening the CO_2 inflow, the pH value slowly decreases. This makes it possible to use a relatively simple controller. In fact, we used a simple, programmable threshold switch, which was sufficiently accurate to control the pH value in the desired interval (7.35 – 7.45, c.f. Fig. 4.3b). At the same time, this concept can be easily combined with various other experimental setups, e.g. electrochemical tests [12] (potentiodynamic and electrochemical impedance measurements in static or rotating electrode setups), determination of released ions [17, 37], use of laminar flow test bench [17, 18, 21, 38], additional control of the O_2 concentration [21], or to determine susceptibility to stress corrosion cracking [39].

4.4.2 Influence of the testing conditions on degradation rates

4.4.2.1 Pure iron

The degradation rate of pure Fe in SBF(CO_2) decreases slightly with prolonged immersion time. This indicates the formation of a layer of degradation products which hinders further degradation [12, 15]. In order to relate the results of this study to previous findings, we considered the interval between 14 and 28 days and determined an average mass loss rate of ≈ 0.22 mg $\cdot cm^{-2} \cdot d^{-1}$), which corresponds to ≈ 0.10 mm $\cdot y^{-1}$. This value is slightly lower than other literature data. (All results were converted to mm $\cdot y^{-1}$ to allow their comparison, even though this requires uniform degradation to take place, which is not always a given [5].) Zhu et al. [38] determined an average value over four weeks of 0.23 mm · y^{-1} in an unspecified SBF. A number of studies are available which used Hank's solution as the testing medium. Liu et al. [21] determined the degradation rate over 30 days to be 0.19 mm · y^{-1}. The value determined by Moravej et al. [37] was 0.14 mm · y^{-1} in Hank's solution buffered using Hepes. All studies used dynamic testing conditions, and are hence better comparable to the present situation than static immersion conditions. These results indicate that degradation in the absence of a pH buffer progresses more slowly than in buffered solutions. However, the difference is rather small compared to the case of Mg degradation.

4.4.2.2 Magnesium alloy WZ21

At first it appears surprising that the degradation rate in SBF(CO_2) is so much lower compared to the puffer bearing SBFs. Usually it is argued that very low degradation rates occur only in unbuffered solutions because the pH value rapidly increases (up to a value of 10.5), which promotes the formation of a stable and protective surface film ($Mg(OH)_2$) [13]. It was concluded that higher concentrations of the pH buffer result in more stable conditions in the testing solution. The slower increase in the pH value is therefore responsible for the higher degradation rates [12, 13]. However, in the CO_2-buffered setup the pH value was always in the range between 7.35 and 7.45, and hence the lowest of all testing solutions investigated (Fig. 4.3). Nevertheless, the degradation rate in SBF(CO_2) is the lowest of all three SBFs. These findings demonstrate the strong influence of additional pH buffers (such as Tris or Hepes) on the degradation behavior of Mg alloys. It was previously reported that adding Hepes or HCO_3^- (as a pH buffer) leads to an increase in the degradation rate of pure Mg [2, 11, 24, 25, 40]. The type of buffer is also important, as shown in Fig. 4.4. Compared to SBF(CO_2), the degradation rate in SBF(Tris) is approximately five times higher, whereas the increase in SBF(Hepes) is about 300 times higher. Even though the concentration of Hepes was twice that of Tris (100 mmol·l^{-1} compared to 50 mmol·l^{-1}), the acceleration of the degradation caused by employing Hepes is greater by far more than twice, in fact by a factor of 60. Hepes (apparent $p_{Ka2} \approx 7.3$ at 37 °C [41]) is much more effective at buffering in the physiological pH range of 7.4 than is Tris (apparent $p_{Ka2} \approx 7.9$ at 37 °C [41]), which has a poor buffering capacity at pH values lower than 7.5 [41].

In the presence of a pH buffer the hydroxide ions produced during Mg degradation are absorbed by the buffer, which has two consequences. On one hand it promotes the forward reaction of the Mg dissolution [2], and on the other fewer hydroxide ions are available for the formation of protective surface films [2, 25, 36]. Moreover, Kirkland et al. [11] suggested an interaction or complex formation between Hepes and Mg^{2+} causing a lower amount of Mg^{2+} available for the formation of surface films.

Usually, carbonates (e.g. $MgCO_3$, especially promoted by a high HCO_3^- concentration) and phosphates form during Mg degradation in SBF [25]. Due to the degradation reaction, the pH value on the sample surface is locally higher compared to the remaining solution. This promotes the formation of the above compounds on the sample surface, where they eventually generate a decrease in the degradation rate due to mass transport limitations through the precipitates [2, 12, 15, 42]. The presence of Hepes in the testing solution was found to affect the composition and kinetics of the surface layers in such a way as less Ca and P were detected on the sample surface compared to a $NaHCO_3/CO_2$ buffered solution [11]. The choice of pH buffer seems to influence the nucleation of degradation products not only by keeping the surface pH value lower, but it was also suggested that a buffer-regulated precipitation has to be considered [11].

The degradation behavior of pure Mg in an unbuffered SBF (with a comparatively low HCO_3^- concentration of 4.2 mmol $\cdot l^{-1}$), a Tris-buffered SBF and additionally in a pure Tris solution was recently investigated [24]. The initial degradation in particular varied significantly between the three testing solutions, and was highest for the Tris solution and slowest for the unbuffered SBF. The degradation in Tris-buffered SBF lay between the other two curves. These results indicate the accelerating effect of the pH buffer and also the decrease in the degradation rate in the presence of HCO_3^-. An increase in the HCO_3^- concentration (from 4.0 to 27 mmol $\cdot l^{-1}$) was also shown to cause a decreased degradation rate of pure Mg in Tris-buffered SBF [36]. At the same time the total buffer concentration was kept constant by reducing the Tris concentration. The authors argued that an increasing HCO_3^- concentration produces a more rapid formation of rather dense degradation products, which limit further degradation. However, they did not take into account the effects of altering the buffering species (i.e. from Tris to HCO_3^-) on degradation behavior. The results from Fig. 4.4 indicate clearly that they have to be considered. In view of the significance of in vitro experiments, we believe the testing setup used here to be superior, as it completely avoids the use of pH buffers.

Proteins are expected to influence the degradation rate of Mg implants during in vitro tests [8]. Indeed, it was found that adding protein to the testing solution decreases the degradation rate [20, 25, 43, 44]. However, in view of the fact that the corresponding mechanism is not yet fully understood [20], it might be of interest to use the CO_2 buffer system in combination with protein-containing solutions to exclude the possible influence of pH changes.

4.4.3 Correlation with in vivo results

One of the most important goals of in vitro tests is the ability to predict behavior in vivo. It is therefore reasonable to mimic the in vivo situation with the aim of deducing reliable predictions from laboratory tests.

An initial in vivo study using WZ21 disks implanted in various tissues in minipigs showed encouraging results [12]. Homogenous degradation and only limited gas formation were observed. The tissue adjacent to the implants was also vascularized, which indicates the uptake of the tissues' physiological functionality. In a subsequent study, the degradation behavior of WZ21 pins implanted in growing femoral rat bones was investigated [6]. The use of μCT allowed monitoring of the implants in the living animals over the entire period of the study. The pin volume, pin surface and gas volume were also quantified and evaluated. These datasets served as a basis for estimating the in vivo degradation rates of WZ21. The degradation of the pins was seen to be fairly slow during the first eight weeks (c.f. Fig. 1a in [6]). A slight increase in the pin volume was even detected after four weeks due to the formation of degradation products surrounding the implant. The loss in pin volume over the first eight weeks served to calculate the lower limiting degradation rate in vivo, as indicated in Fig. 4.3. We assumed that the degradation process was constant after four weeks and used the loss in pin volume up to week eight to calculate an averaged or steady-state degradation rate in vivo. After the eight weeks in vivo, the tendency for localized corrosion increases, which can be seen in the 3-D reconstruction of the pins and deduced from the corresponding increase in pin surface. After twelve weeks in particular the pins show pronounced localized degradation. It is therefore not reasonable to calculate average degradation rates based on the data after week twelve. However, in order to estimate a maximum degradation rate, we used the loss in pin volume from week eight to week twelve. The occurrence of localized degradation generates a pronounced increase in the degradation rate, which is apparent from the in vivo data. It is important to be aware that the above-mentioned estimate results in average degradation rates. In vivo, however, the degradation rate depends strongly on the location in the body. From the in vivo studies it can be deduced that the degradation is influenced by the degree of vascularization of the surrounding tissue [3, 6, 12, 45, 46]. A higher degree of vascularization, e.g. in the medullary cavity, promotes the degradation compared to a lower degree of vascularization, as e.g. in cortical bone. It is hence necessary to specify the exact location of the implant in the body. When considering the average degradation in vivo (from week four to week eight) the resulting rate agrees well with the in vitro degradation rate in $SBF(CO_2)$. The approach used in this study therefore appears suitable for assessing in vivo degradation performance, at least for well-vascularized locations in living organisms.

4.5 Conclusions

Reliable information on the degradation behavior in vivo gained from laboratory experiments is the basis for the successful development of biodegradable metals. In this study, gaseous CO_2 was used to control the pH value of the immersion testing medium. It hence uses the same mechanism as the human body to maintain the pH value in a narrow interval around 7.40 ± 0.05. By avoiding pH buffers, the experimental conditions also mimic the physiological situation much more closely. It was shown that the use of pH buffers strongly accelerates the degradation rate and that pH values increase readily during immersion. Both issues are circumvented by the approach used in this study, which may be applied to investigate both biodegradable Fe and Mg alloys. Owing to its modular design, the approach can also be easily combined with other setups, e.g. laminar flow cells, electrochemical tests, or used to determine solute ion concentrations. The degradation rate found for the Mg alloy WZ21 agrees well with results from the in vivo study, which demonstrates the approach's potential.

Acknowledgements

The authors very much appreciate financial support received within the framework of the project "Biocompatible Materials and Applications" initiated by the Austrian Institute of Technology GmbH (AIT), and support from the Staub/Kaiser Foundation, Switzerland.

References

[1] Waksman R, Pakala R, Baffour R, Seabron R, Hellinga D, Tio FO. Short-term effects of biocorrodible iron stents in porcine coronary arteries. J Interv Cardiol 2008;21:15-20.

[2] Xin Y, Hu T, Chu PK. In vitro studies of biomedical magnesium alloys in a simulated physiological environment: A review. Acta Biomater 2011;7:1452-9.

[3] Witte F, Hort N, Vogt C, Cohen S, Kainer KU, Willumeit R, et al. Degradable biomaterials based on magnesium corrosion. Curr Opin Solid State Mater Sci 2009;12:63-72.

[4] Hermawan H, Dubé D, Mantovani D. Developments in metallic biodegradable stents. Acta Biomater 2010;6:1693-7.

[5] Kirkland NT, Birbilis N, Staiger MP. Assessing the corrosion of biodegradable magnesium implants: A critical review of current methodologies and their limitations. Acta Biomater 2012;8:925-36.

[6] Kraus T, Fischerauer SF, Hänzi AC, Uggowitzer PJ, Löffler JF, Weinberg AM. Magnesium alloys for temporary implants in osteosynthesis: In vivo studies of their degradation and interaction with bone. Acta Biomater 2012;8:1230-8.

[7] Moravej M, Mantovani D. Biodegradable Metals for Cardiovascular Stent Application: Interests and New Opportunities. Int J Mol Sci 2011;12:4250-70.

[8] Virtanen S. Biodegradable Mg and Mg alloys: Corrosion and biocompatibility. Mater Sci Eng B 2011;176:1600-8.

[9] Peuster M, Hesse C, Schloo T, Fink C, Beerbaum P, von Schnakenburg C. Long-term biocompatibility of a corrodible peripheral iron stent in the porcine descending aorta. Biomaterials 2006;27:4955-62.

[10] Witte F, Fischer J, Nellesen J, Crostack HA, Kaese V, Pisch A, et al. In vitro and in vivo corrosion measurements of magnesium alloys. Biomaterials 2006;27:1013-8.

[11] Kirkland N, Waterman J, Birbilis N, Dias G, Woodfield T, Hartshorn R, et al. Buffer-regulated biocorrosion of pure magnesium. J Mater Sci: Mater Med 2012;23:283-91.

[12] Hänzi AC, Gerber I, Schinhammer M, Löffler JF, Uggowitzer PJ. On the in vitro and in vivo degradation performance and biological response of new biodegradable Mg–Y–Zn alloys. Acta Biomater 2010;6:1824-33.

[13] Zainal Abidin NI, Atrens AD, Martin D, Atrens A. Corrosion of high purity Mg, Mg2Zn0.2Mn, ZE41 and AZ91 in Hank's solution at 37 °C. Corros Sci 2011;53:3542-56.

[14] Gu X, Zheng Y, Cheng Y, Zhong S, Xi T. In vitro corrosion and biocompatibility of binary magnesium alloys. Biomaterials 2009;30:484-98.

[15] Hänzi AC, Gunde P, Schinhammer M, Uggowitzer PJ. On the biodegradation performance of an Mg–Y–RE alloy with various surface conditions in simulated body fluid. Acta Biomater 2009;5:162-71.

[16] Hermawan H, Dube D, Mantovani D. Degradable metallic biomaterials: Design and development of Fe–Mn alloys for stents. J Biomed Mater Res, Part A 2010;93A:1-11.

[17] Hermawan H, Purnama A, Dube D, Couet J, Mantovani D. Fe–Mn alloys for metallic biodegradable stents: Degradation and cell viability studies. Acta Biomater 2010;6:1852-60.

[18] Lévesque J, Hermawan H, Dubé D, Mantovani D. Design of a pseudo-physiological test bench specific to the development of biodegradable metallic biomaterials. Acta Biomater 2008;4:284-95.

[19] Li Z, Gu X, Lou S, Zheng Y. The development of binary Mg–Ca alloys for use as biodegradable materials within bone. Biomaterials 2008;29:1329-44.

[20] Willumeit R, Fischer J, Feyerabend F, Hort N, Bismayer U, Heidrich S, et al. Chemical surface alteration of biodegradable magnesium exposed to corrosion media. Acta Biomater 2011;7:2704-15.

[21] Liu B, Zheng YF. Effects of alloying elements (Mn, Co, Al, W, Sn, B, C and S) on biodegradability and in vitro biocompatibility of pure iron. Acta Biomater 2011;7:1407-20.

[22] Greenbaum J, Nirmalan M. Acid-base balance: The traditional approach. Curr Anaesth Crit Care 2005;16:137-42.

[23] Seitz J-M, Collier K, Wulf E, Bormann D, Bach F-W. Comparison of the Corrosion Behavior of Coated and Uncoated Magnesium Alloys in an In Vitro Corrosion Environment. Adv Eng Mater 2011;13:B313-B23.

[24] Xin Y, Chu PK. Influence of Tris in simulated body fluid on degradation behavior of pure magnesium. Mater Chem Phys 2010;124:33-5.

[25] Yamamoto A, Hiromoto S. Effect of inorganic salts, amino acids and proteins on the degradation of pure magnesium in vitro. Mater Sci Eng C 2009;29:1559-68.

[26] Fischer J, Pröfrock D, Hort N, Willumeit R, Feyerabend F. Improved cytotoxicity testing of magnesium materials. Mater Sci Eng B 2011;176:830-4.

[27] Zhang S, Li J, Song Y, Zhao C, Zhang X, Xie C, et al. In vitro degradation, hemolysis and MC3T3-E1 cell adhesion of biodegradable Mg–Zn alloy. Mater Sci Eng C 2009;29:1907-12.

[28] Zhang S, Zhang X, Zhao C, Li J, Song Y, Xie C, et al. Research on an Mg–Zn alloy as a degradable biomaterial. Acta Biomater 2010;6:626-40.

[29] Atherton JC. Acid-base balance: maintenance of plasma pH. Anaesth Intensive Care Med 2009;10:557-61.

[30] Hänzi AC, Sologubenko AS, Uggowitzer PJ. Design strategy for new biodegradable Mg–Y–Zn alloys for medical applications. Int J Mater Res 2009;100:1127-36.

[31] Hänzi AC, Metlar A, Schinhammer M, Aguib H, Lüth TC, Löffler JF, et al. Biodegradable wound-closing devices for gastrointestinal interventions: Degradation performance of the magnesium tip. Mater Sci Eng C 2011;31:1098-103.

[32] Song GL, Atrens A, StJohn D. An hydrogen evolution method for the estimation of the corrosion rate of magnesium alloys. In: Hryn JN, editor. Magnesium Technology 2001. New Orleans, LA: The Minerals, Metals & Materials Society; 2001. p. 255-62.

[33] Müller L, Müller FA. Preparation of SBF with different HCO_3^- content and its influence on the composition of biomimetic apatites. Acta Biomater 2006;2:181-9.

[34] Moretti G, Guidi F, Grion G. Tryptamine as a green iron corrosion inhibitor in 0.5 M deaerated sulphuric acid. Corros Sci 2004;46:387-403.

[35] Zhu PX, Masuda Y, Yonezawa T, Koumoto K. Investigation of apatite deposition onto charged surfaces in aqueous solutions using a quartz-crystal microbalance. J Am Ceram Soc 2003;86:782-90.

[36] Xin Y, Hu T, Chu PK. Degradation behaviour of pure magnesium in simulated body fluids with different concentrations of HCO_3^-. Corros Sci 2011;53:1522-8.

[37] Moravej M, Purnama A, Fiset M, Couet J, Mantovani D. Electroformed pure iron as a new biomaterial for degradable stents: In vitro degradation and preliminary cell viability studies. Acta Biomater 2010;6:1843-51.

[38] Zhu S, Huang N, Xu L, Zhang Y, Liu H, Sun H, et al. Biocompatibility of pure iron: In vitro assessment of degradation kinetics and cytotoxicity on endothelial cells. Mater Sci Eng C 2009;29:1589-92.

[39] Choudhary L, Singh Raman RK. Magnesium alloys as body implants: Fracture mechanism under dynamic and static loadings in a physiological environment. Acta Biomater 2012;8:916-23.

[40] Quach NC, Uggowitzer PJ, Schmutz P. Corrosion behaviour of an Mg–Y–RE alloy used in biomedical applications studied by electrochemical techniques. C R Chim 2008;11:1043-54.

[41] Good NE, Winget GD, Winter W, Connolly TN, Izawa S, Singh RMM. Hydrogen Ion Buffers for Biological Research. Biochem 1966;5:467-77.

[42] Lindström R, Johansson L-G, Thompson GE, Skeldon P, Svensson J-E. Corrosion of magnesium in humid air. Corros Sci 2004;46:1141-58.

[43] Liu C, Xin Y, Tian X, Chu PK. Degradation susceptibility of surgical magnesium alloy in artificial biological fluid containing albumin. J Mater Res 2007;22:1806-14.

[44] Rettig R, Virtanen S. Time-dependent electrochemical characterization of the corrosion of a magnesium rare-earth alloy in simulated body fluids. J Biomed Mater Res, Part A 2008;85A:167-75.

[45] Xu L, Yu G, Zhang E, Pan F, Yang K. In vivo corrosion behavior of Mg–Mn–Zn alloy for bone implant application. J Biomed Mater Res, Part A 2007;83A:703-11.

[46] Willbold E, Kaya AA, Kaya RA, Beckmann F, Witte F. Corrosion of magnesium alloy AZ31 screws is dependent on the implantation site. Mater Sci Eng B 2011;176:1835-40.

5 Degradation properties II - Degradation performance of Fe–Mn–C(–Pd) alloys

The TWIP alloys are attractive for their good mechanical performance. Based on the considerations presented in Chapter 4, Chapter 5 investigates their degradation properties in detail, primarily via immersion testing and electrochemical impedance spectroscopy. From these experiments important conclusions were drawn regarding the suitability of Fe-based alloys for application in degradable implants.

Degradation performance of biodegradable Fe–Mn–C(–Pd) alloys[1]

Biodegradable metals offer great potential in circumventing the long-term risks and side effects of medical implants. Austenitic Fe–Mn–C–Pd alloys feature a well-balanced combination of high strength and considerable ductility which make them attractive for use as degradable implant material. The focus of this study is the evaluation of the degradation performance of these alloys by means of immersion testing and electrochemical impedance spectroscopy in simulated body fluid. The Fe–Mn–C–Pd alloys are characterized by an increased degradation rate compared to pure Fe, as revealed by both techniques. Electrochemical measurements turned out to be a sensitive tool for investigating the degradation behavior. They not only show that the polarization resistance is a measure of corrosion tendency, but also provide information on the evolution of the degradation product layers. The mass loss data from immersion tests indicate a decreasing degradation rate for longer times due to the formation of degradation products on the sample surfaces. The results are discussed in detail in terms of the degradation mechanism of Fe-based alloys in physiological media.

5.1 Introduction

The development of biodegradable metals for temporary medical implants has been the subject of intense research in recent years [1-3]. The potential applications of these metals include not only osteosynthesis and coronary stents, but also others, such as degradable wound closing devices [4]. Degradable implants require only one intervention and eliminate the need to remove the implant in a second operation. Moreover, they potentially reduce long-term risks and side effects, e.g. chronic inflammation, in-stent restenosis or the inability to adapt to the growing blood vessel (in the case of stents) [5-7]. The suitability of iron as

---

[1]M. Schinhammer, P. Steiger, F. Moszner, J.F. Löffler, P.J. Uggowitzer; Materials Science and Engineering: C 33 (2013) 1882-1893

5.1 Introduction

degradable implant material has been shown by in vivo studies, where stents made of pure iron were investigated [5, 7, 8]. No indications of local or systemic toxicity, no local inflammations, and no early restenosis due to thrombotic processes were detected. However, the degradation rate of pure iron in vivo was found to be too low, approaching the behavior of permanent implant materials [5, 7]. In addition, the mechanical properties of pure iron are not well suited for use as implant material.

In the following, Hermawan et al. [9, 10] presented Fe–Mn alloys which feature increased degradation rates and mechanical properties similar to those of stainless steel 316L. To further explore the possibilities of biodegradable Fe-based alloys, we developed a design strategy to achieve both higher degradation rates and superior mechanical performance [11]. It consists of a controlled modification of the chemical composition and the microstructure, and relies on two factors: (i) adding Mn lowers the standard electrode potential of the matrix; and (ii) the formation of noble Pd-rich precipitates is expected to induce microgalvanic corrosion, which greatly enhances the degradation rate [11]. Additionally, the controlled precipitation reaction can be used to specifically influence the microstructure and mechanical properties, as shown for martensitic Fe–Mn–Pd [12] and austenitic Fe–Mn–C–Pd alloys [13]. The latter alloys, which are investigated in the present study, combine the high strength of Co–Cr–Mo alloys with the ductility of stainless steel 316L, and thus offer better performance than these commonly used materials [13].

Apart from microstructure and mechanical properties, degradation behavior is of great importance for intended use of pure Fe and biodegradable Fe-based alloys as degradable implant material. The majority of the studies on these alloys have been performed in simulated physiological fluids, using either simulated body fluid (SBF) [11, 14] or Hank's balanced salt solution [9, 10, 15-18] to reproduce in vivo conditions. Different testing setups were also used, which included static immersion tests [10, 17-19], dynamic immersion tests in a laminar flow test bench [14-16, 19], (static) electrochemical measurements [9, 10, 17-20], and determination of the ion concentrations in solution [15-17, 19, 20]. It was previously shown for Mg-based alloys that the composition of the solution (ion and protein concentrations) and in particular the type of buffering agent strongly influence degradation behavior [21-28]. Consequently, it was suggested to use gaseous CO_2 to maintain the testing solutions at constant pH [26]. Active regulation was also recently suggested for even tighter regulation of the pH value [21]. It was shown that the degradation rates obtained using this approach correlate rather well with results from in vivo studies on well-vascularized tissues [21]. Based on these results, we used this setup to assess the degradation behavior of newly developed Fe–Mn–C–Pd alloys, which we investigated by means of immersion tests and electrochemical impedance spectroscopy (EIS). We then analyzed the degradation products and measured the elemental distribution. The implications of our findings for the potential use of Fe–Mn–C–Pd alloys as degradable implant material are also discussed in detail in this work.

5.2 Materials and Methods

5.2.1 Materials

Pure Fe (Armco quality) and two austenitic Fe-based alloys of nominal composition Fe-21Mn-0.7C (designated as TWIP) and Fe-21Mn-0.7C-1Pd (in wt.%, designated as TWIP-1Pd) were investigated in this study. Production and characterization in terms of microstruc-ture and mechanical performance of the TWIP(-1Pd) alloys is given in Ref. [13]. The TWIP alloy was used in a recrystallized (rexx) condition (cold-working of 30%, annealing for 30 min at 900 °C), whereas the TWIP-1Pd alloy was used in various heat treatment conditions. The recrystallized condition (cold-working of 56%, annealing for 10 min at 1150 °C) was used as a reference state. In addition, some samples were subsequently cold-worked (swaged) to a cold-working degree of either 12% or 23% and annealed for 30 min at either 700 °C or 900 °C. Their designation is composed of the degree of cold-working and the annealing temperature, e.g. TWIP-1Pd CW12-700C. During the annealing, strain-induced heterogeneous precipitation of Pd-rich particles and recovery occurred. The heat treatment states designated as TWIP-1Pd CW12-700C and TWIP-1Pd CW23-700C were chosen in particular because of their advantageous combination of strength and ductility [13]. The heat treatment state designated as TWIP-1Pd CW23-900C was also investigated. Here partial recrystallization was observed; however, the main difference that arises from annealing at the higher temperature is the size of the Pd-rich precipitates formed [13]. These samples were investigated to evaluate the influence of the precipitates' size on the degradation rate.

5.2.2 Methods

5.2.2.1 Microstructure characterization

Light optical microscopy on corroded samples was performed using a stereo microscope (Leica MZ 12.5) and an optical microscope (Reichert-Jung Polyvar met).

Cross-sections of samples immersed for 14 d in SBF were prepared as follows: after immersion, the samples were dried at 40 °C in air for 72 h. They were then embedded in electrically conductive cold mounting resin (Demotec 70, Demotec). Subsequent preparation consisted of grinding (up to P1200 grit SiC paper) and polishing (up to 1 μm diamond polish) with careful cleaning between each preparation step using cotton wool and ethanol. Final polishing was done using a 0.05 μm Al_2O_3 suspension (MasterPrep, Buehler). Elemental distribution maps of the degradation products were recorded using a Hitachi SU-70 scanning electron microscope (SEM, Schottky-type field emission gun, operating at 20 kV acceleration voltage) equipped with an X-max energy dispersive X-ray (EDX) detector (Oxford instruments).

Additional microstructure characterization was performed by electron backscattered diffraction (EBSD) in the SEM, which was further equipped with a Nordlys EBSD camera (Oxford Instruments). The EBSD scans were recorded using 20 kV acceleration voltage and a probe current of approximately 2 nA at 17 mm working distance. The samples were embedded in bakelite and ground as well as polished down to 0.25 μm diamond size, and the final polishing step was carried out using colloidal silica suspension (Buehler MasterMet 2). The indexing rate for the plots shown was generally above 95%.

Transmission electron microscopy (TEM) was performed on a FEI Tecnai F30 machine operated at 300 kV. The atomic-number-sensitive high-angle-annular dark-field (HAADF) imaging mode of the scanning TEM (STEM) was deployed to show the compositional con-trast. The elemental distribution was determined using an energy dispersive X-ray (EDX) detector in the STEM mode. The TEM specimens were prepared by mechanically grinding the Fe-based samples to a thickness of approximately 100 μm. Disks of 3 mm in diameter were punched out from these specimens. These disks were then dimpled on one or both sides using a Gatan dimple grinder with 1 μm diamond suspension (Metadi oil-based). Electron transparency was obtained by twin-jet electro polishing (Tenu-Pol 5, Struers) using 24 V DC at a temperature of –30 °C, using a solution of 5 vol.% perchloric acid in methanol as electrolyte.

Atom probe tomography (APT) measurements were performed on a Leap 4000X HR (Cameca) in the voltage pulse mode (pulse fraction of 15%) under ultra-high vacuum ($< 10^{-10}$ mbar) conditions and a sample temperature of 80 K. Data reconstruction and statistical evaluation was conducted using the IVAS 3.4.1 software package (Cameca). Square bars of 0.3 x 0.3 x 20 mm^3 were cut from the rods by spark erosion and then etched to sharp needle-like specimens by a standard two-step electropolishing procedure [29]. The first solution consisted of 10 vol.% perchloric acid in methanol, followed by polishing in 2 vol.% perchloric acid in butoxyethanol.

5.2.2.2 Immersion testing

For immersion tests, 3.5 mm-thick disks of 8 mm (TWIP, TWIP-1Pd rexx), and 7 mm (pure Fe, TWIP-1Pd cold-worked and annealed) in diameter were prepared. The entire surface was ground up to P1200 grit SiC paper, cleaned in ethanol in an ultrasonic bath and dried in hot air.

Immersion testing was performed in two different environments: in SBF at 37 $\pm$ 1 °C, and in 0.5 M H_2SO_4 at room temperature. The immersion testing under near-physiological conditions in SBF was carried out as described previously [21]: the samples were immersed in SBF for different times of up to one month.

The pH value was regulated using gaseous CO_2, which was fed into the testing solution. By this means the pH was kept more or less constant in the interval between 7.35 and 7.45 and it was not necessary to add an additional pH-buffer. A sufficient amount of SBF (composition given in Table 1 of Ref. [30], and in Table 4.1 on page 89) was deployed to ensure a high solution-volume-to-sample-surface ratio [31], which in this study was 160 ml·cm^{-2}. After immersion the samples were mechanically cleaned; the remaining degradation products were chemically removed; and the mass loss per surface area was determined. Consequently each measurement point was established independently using freshly prepared samples, because a re-immersion of the cleaned samples would have biased the results.

Immersion in H_2SO_4 was performed in order to investigate the possible influence of the size of the precipitates on degradation rate. In the acidic environment, hardly any degradation products able to limit the diffusion of the reacting species are formed on the sample surface. In addition, hydrogen evolution is the dominant cathodic reaction, allowing the degradation rate to be measured via the amount of evolved hydrogen, as previously described in the context of Mg-alloys [32]. The experimental setup outlined in Ref. [33] was used: the samples were placed in a beaker containing H_2SO_4, and a measuring cylinder (also filled with H_2SO_4) was placed over the samples to collect the hydrogen gas which formed during the experiment. The amount of hydrogen collected in the measuring cylinder makes it possible to measure the corrosion rate over time.

5.2.2.3 Electrochemical measurements

EIS measurements were recorded with the NOVA 1.8 software package using an Autolab PGSTAT302 device with an additional frequency response analysis module (FRA2, Eco Chemie B.V.). The measured frequencies ranged between 10^5 and $7 \cdot 10^{-2}$ Hz in the single sine acquisition mode with an applied amplitude of 10 mV. The measurements were recorded at the open circuit potential (OCP), which was initially allowed to equilibrate for 1 h. Thereafter, EIS spectra were recorded in regular intervals.

5 Degradation properties II - Degradation performance of Fe–Mn–C(–Pd) alloys

The samples were measured with a rotating disc electrode (RDE) setup (Pine Research Instruments), where the sample was the rotating working electrode (rotation rate of 500 min^{-1}). A saturated calomel electrode (KCl / Hg_2Cl_2) served as a reference electrode, and a flat, rectangular platinum sheet as a counter electrode. Cylindrical samples of 5 mm in diameter and 4 mm in length were inserted into a sample holder made from polytetrafluoroethylene (PTFE). The outer edge of the sample was sealed with silicone (Dow Corning 732) to prevent crevice corrosion between the sample and the holder. The effective surface (approximately 0.13 cm^2) exposed to the electrolyte was determined for each sample individually. The experimental setup is shown schematically in Fig. 5.1. As electrolyte, 250 ml of SBF heated to 37 ± 1 °C was used. Generally, the pH value was

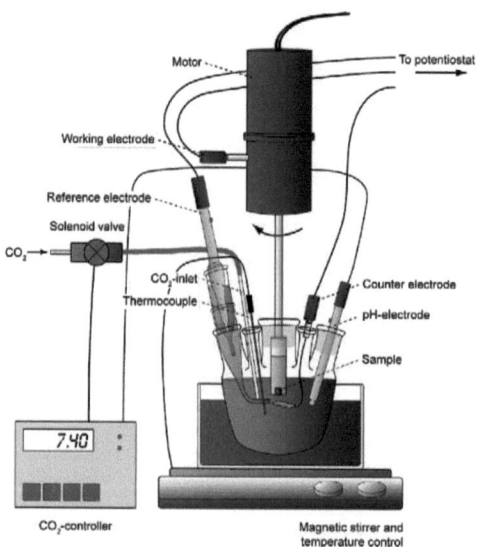

Figure 5.1: Schematic illustration of the testing setup used for electrochemical measurements. The RDE setup is combined with pH regulation using CO_2: The pH value is monitored by a programmable threshold switch that opens a solenoid valve if the pH value exceeds the predefined upper threshold value. Gaseous CO_2 is then fed into the SBF until the lower threshold is reached. The sample is rotated to generate the flow of the SBF. A magnetic stirrer is used to control SBF temperature.

controlled as described above, using gaseous CO_2. To investigate the possible influence of a pH buffer on the degradation behavior, additional experiments were conducted in SBF containing 100 mmol·l^{-1} Hepes (Carl Roth GmbH). This solution is referred to as SBF(Hepes).

5.3 Results

5.3.1 Microstructure characterization

In the reference states (i.e. the recrystallized state) both the TWIP and TWIP-1Pd alloys possess a regular microstructure consisting of equiaxed grains which contain some annealing twins [13]. The corresponding grain sizes are approximately 40 μm and 150 μm for TWIP and TWIP-1Pd, respectively. The samples differ in their grain size due to the higher annealing temperature employed for TWIP-1Pd [13]. For the TWIP-1Pd samples it is also important to note that no Pd-rich precipitates are present in the matrix [13].

The microstructures of the TWIP-1Pd samples in the heat treatment states CW23-700C and CW23-900C are presented in Fig. 5.2. A detailed description of the microstructural features and mechanical performance of the same alloy in the different heat treatment states is given in Ref. [13]. The microstructures in both CW23-700C (Fig. 5.2a) and CW23-900C (Fig. 5.2d) states appear similar from the EBSD maps. The grains contain a large number of deformation twins. However, these two heat treatment states differ in the size of the Pd-rich precipitates present in the alloys in particular. During annealing at 700 °C, relatively small precipitates form of approximately 4 nm in size. The atomic positions of Fe, Mn, C and Pd are shown in the three-dimensional (3D) reconstruction of the APT data (Fig. 5.2b). Pd-enriched regions are already visible from the corresponding atom maps. In order to better visualize the precipitates, isoconcentration surfaces of Pd (at 5 at.%) and Mn (at 26 at.%) are plotted in Fig. 5.2c. They are used to show regions of different composition and reveal that the precipitates are enriched not only in Pd but also in Mn. Based on the here-used reconstruction parameters the precipitates seem plate-shaped. A proximity histogram (Fig. 5.2c) based on the Pd-isoconcentration surface was calculated to obtain information on the precipitate composition. Increasing distance values correspond to the inside of the isoconcentration surfaces and consequently also the precipitates. Clearly the Mn and Pd concentrations are increased in the precipitate, while the Fe concentration is significantly lower. The C concentration also decreases within the precipitates, although this is difficult to see from the picture.

Annealing at 900 °C generates elongated Pd-rich precipitates approximately 20-50 nm in size, as shown in the STEM image (acquired in [0 1 1] zone-axis orientation) in Fig. 5.2e. The Pd-rich precipitates appear with a bright contrast in the atomic-number-sensitive HAADF signal. Lines with bright contrast also indicate the decoration of lattice defects with Pd. EDX mapping was performed on the area indicated in Fig. 5.2e, and the Pd $L_{\alpha 1}$ and Fe $K_{\alpha 1}$ elemental distribution maps are shown in Fig. 5.2f. The intensities of the EDX signal along the lines indicated in the Pd map are given in the plots in Fig. 5.2f below. The line profile 1 confirms the visual impression from the elemental maps that the precipitates are enriched in Pd (and Mn) and depleted in Fe. The line profile 2 across a dislocation indicates an enrichment of Pd in the dislocation core running perpendicular to the line profile.

5.3.2 Immersion testing in SBF

Immersion testing revealed distinct differences in the degradation performance of the alloys. Fig. 5.3a shows the mass loss of Fe, TWIP, and two TWIP-1Pd alloys as a function of immersion time in SBF. The samples containing Pd feature the highest mass loss, irrespective of their heat treatment state. The degradation of the TWIP samples proceeds more slowly and the slowest degradation was found for pure Fe. The corresponding mass loss rates are plotted in Fig. 5.3b. Common to all samples is that the degradation rate decreases significantly with prolonged immersion time. From an application point of view, the steady state degradation rate (i.e. the degradation rate > 14 d) is of particular interest.

5.3 Results

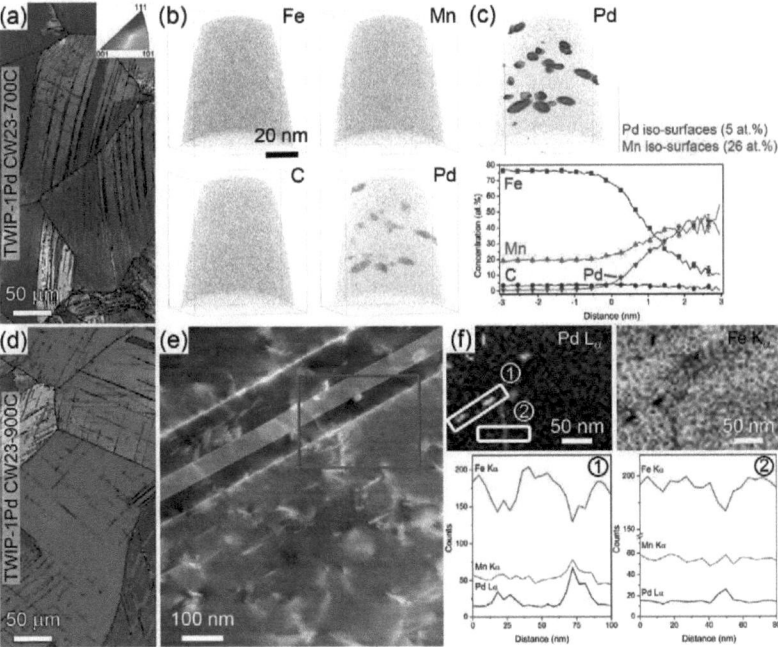

Figure 5.2: Microstructure characterization of TWIP-1Pd CW23-700C (a-c) and TWIP-1Pd CW23-900C (d-f): (a) EBSD map (IPF out-of-plane color coding) showing an overview of the microstructure. High-angle grain boundaries are indicated by solid black lines, and solid red lines correspond to 60° ⟨1 1 1⟩-twin boundaries. (b) 3D reconstruction showing atom maps of Fe (blue), Mn (red), C (purple), and Pd (green). (c) Both Pd (5 at.%) and Mn (26 at.%) isoconcentration surfaces and the proximity histogram reveal the enrichment of precipitates with Pd and Mn. (d) EBSD map (IPF out-of-plane color coding) of the microstructure. (e) The HAADF-STEM image ([0 1 1] zone-axis orientation) shows that the precipitates and the lattice defects are enriched in Pd (heaviest alloy constituent). The area of the EDX mapping is also indicated. (f) Pd $L_{\alpha 1}$ and Fe $K_{\alpha 1}$ elemental distribution maps, and line profile data over two precipitates (1) and over a dislocation line (2). These show that the precipitates and lattice defects are Pd-enriched.

117

5 Degradation properties II - Degradation performance of Fe–Mn–C(–Pd) alloys

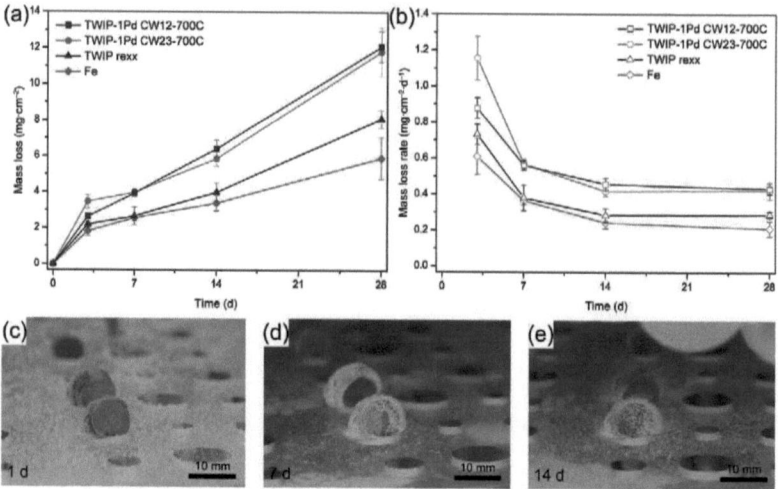

Figure 5.3: Results from immersion testing in SBF. (a) Mass loss as a function of immersion time. The TWIP-1Pd samples show the highest mass loss. (b) Mass loss rate for the data shown in (a). The TWIP-1Pd samples feature the highest mass loss rates. Characteristic of all samples is the continuous decrease in the mass loss rate for prolonged immersion time. Images of TWIP rexx after (c) 1 d; (d) 7 d; and (e) 14 d of immersion.

The photographs in Fig. 5.3c-e show an example degradation process involving TWIP samples over 14 d. Even after just one day the sample is partially covered with brownish-white degradation products. It is worth noting that the edges of the samples were often attacked first. With prolonged immersion time degradation proceeds and spreads over the entire sample surface; from the photographs it is evident that it coincides with the appearance of voluminous degradation products.

5.3.3 Cross-sections of immersed samples

The images in Fig. 5.4 show the morphology and composition of cross-sections after immersion in SBF for 14 d. In both samples, TWIP rexx (Fig. 5.4a) and TWIP-1Pd CW23-700C (Fig. 5.4b), a layered degradation product structure is already apparent from the backscattered electron (BSE) contrast (left images). The

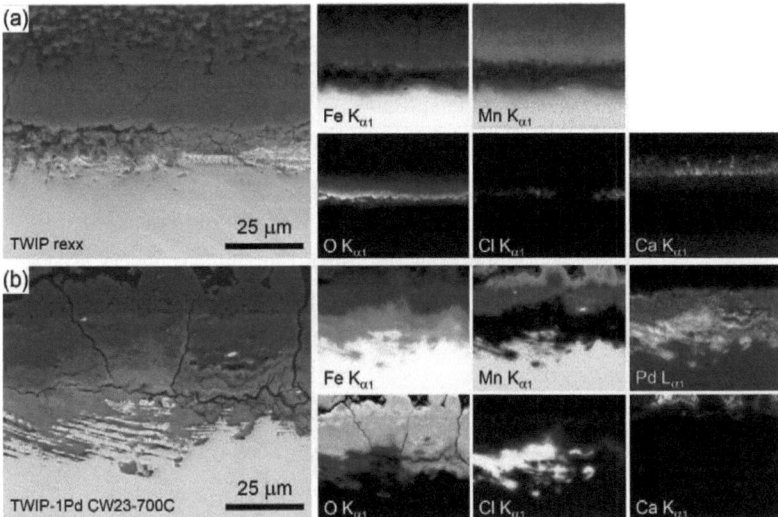

Figure 5.4: Cross-sections of (a) TWIP rexx and (b) TWIP-1Pd CW23-700C after immersion in SBF for 14 d. The images show the backscattered electron contrast (left) and the corresponding elemental distribution maps (right). The EDX maps reveal that the degradation products consist of layers with different compositions: the layer adjacent to the metal surface contains mainly Fe, Cl, O, and Pd (in the case of TWIP-1Pd), whereas the next layer is composed of Fe- and Mn-oxides. Finally, the Ca-rich layer was formed from depositions from the SBF.

EDX maps (right-hand images) also indicate compositional differences between the layers.

In the case of TWIP rexx (Fig. 5.4a), the degradation products on top of the metal matrix are structured into three layers. The first layer (directly on top of the metal) appears porous from the BSE contrast. EDX measurements show that it is mainly composed of Fe oxide and contains only small amounts of Mn. Some regions, however, contain significant amounts of Cl, as indicated in the corresponding EDX map. The second layer appears to be denser and is composed of a mixture of Fe and Mn oxides with additional minor amounts of Ca and P (P distribution not shown). The third layer again has a more porous structure and is characterized by a significant amount of Ca and P. Because these elements are only present in the SBF, it may be assumed that the third layer comprises mainly precipitates from the SBF, even though some Fe and Mn were still detected.

The structure of the degradation products on the TWIP-1Pd sample (Fig. 5.4b) is in principal similar to that on TWIP. The different layers can be distinguished even in the BSE contrast. The first layer is mainly composed of Fe oxides and Fe chlorides, plus relatively low amounts of Mn. The Pd distribution is of particular interest, as the corresponding EDX map shows Pd enrichment. In fact, according to EDX measurements the Pd content in this layer (of approximately 2.4 wt.%) is even higher than the bulk concentration (approx. 1.3 wt.%). The consecutive layer of degradation products consists of Fe and Mn oxides with minor amounts of Pd. The third layer features a distinctly different morphology than the first two layers. Because it contains significant concentrations of Ca, P and Na (P and Na distributions not shown) besides Fe, Mn and O, it was concluded that this layer was formed by depositions which originate from SBF.

The corrosion attack generally appears to proceed homogeneously. However, for TWIP-1Pd a comparison with the microstructure reveals the twin boundaries to be preferred degradation sites.

5.3.4 Electrochemical impedance spectroscopy in SBF

The data in Fig. 5.5a show a representative series of EIS spectra from an Fe sample. The Bode representation depicts the impedance (Z modulus) and negative phase angle (φ) as a function of the applied frequency over a period of 16 h. From the phase angle it is apparent that only one time constant (maximum phase angle at approximately 4 Hz) is present in the frequency range investigated. Hence the curves were fitted using an equivalent circuit consisting of a resistor (R_s) connected in series with a parallel combination of a resistor (R_p) and a constant phase element (Q_{dl}), as schematically shown in Fig. 5.5c. The quality of the fits was generally good, and for reasons of clarity the fitted curves are not shown in Fig. 5.5a and b. At high frequencies ($\approx 3 \cdot 10^4$ Hz) the electrochemical double layer is not charged and the system resistance is given by the solution resistance (R_s) only. The increase of the phase angle at 10^5 Hz is an artifact arising from the

5.3 Results

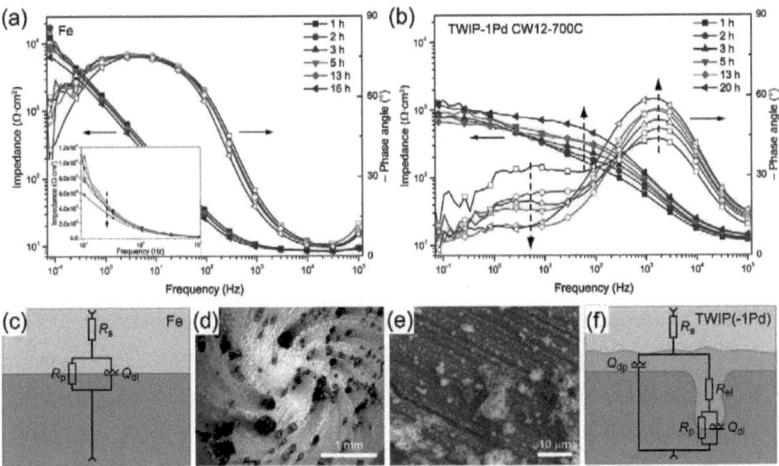

Figure 5.5: EIS spectra of (a) Fe and (b) TWIP-1Pd CW12-700C. The Bode representations show the impedance and the phase angle as a function of the frequency for different measurement times up to 20 h. The inset in (a) shows the low-frequency ends of the Fe spectra. (c) The EIS spectra of Fe contain only one time constant, which was taken into account in the equivalent circuit. (d) Light optical microscopy image of TWIP-1Pd CW12-700C after 24 h in SBF. The surface features localized attacks and degradation products. (e) The SEM image of the surface reveals that the entire sample is covered with a layer of degradation products. (f) Corresponding equivalent circuit for interpreting the results of the TWIP(-1Pd) samples.

electrodes [34]. The low-frequency end of the graph is mainly determined by the polarization (charge transfer) resistance (R_p). The electrochemical double layer which forms at the interface between sample and electrolyte is represented by Q_{dl}. A constant phase element was used to account for surface roughness and heterogeneities [35-38]. The low-frequency parts (< 1 Hz) of the measurements are often disturbed due to active degradation, which interferes with the measurement [31, 34]. However, a close inspection of the low-frequency end of the impedance curves (inset in Fig. 5.5a) reveals a slight decrease in the polarization resistance with increasing measurement time.

121

The EIS spectra of TWIP and TWIP-1Pd samples are clearly different to those of the Fe samples, as illustrated using TWIP-1Pd CW12-700C results (Fig. 5.5b) as examples. The phase shift is evidence of two distinctly separated time constants. As described above, the low-frequency process may be related to the charge transfer reaction. The images of the sample after 24 h in SBF (Figs. 5.5d and e) help to identify the second process at high frequencies (approximately 2 kHz). The light optical image (Fig. 5.5d) shows an overview of the sample. The corrosion attack is localized and pits nucleate at the surface and subsequently grow wider and deeper. Degradation products were formed predominantly in the pits and partially transported out of them due to sample rotation. This generates characteristic "traces" of degradation products, resembling shooting stars. Between the pits it appears that the bare metal surface remains unchanged. However, a close inspection of the surface in the SEM (Fig. 5.5e) reveals that the entire surface is in fact covered with at least a thin layer of degradation products with the characteristic "cracked earth" appearance produced by post-experiment dehydration. Based on the visual inspection of the tested samples, the equivalent circuit schematically shown in Fig. 5.5f was consequently established. In addition to R_s, R_p and Q_{dl} (which have the same meaning as above), a constant phase element (Q_{dp}) and an electrolyte resistance (R_{el}) were introduced which represent the layer of degradation products and the resistance in the pit.

During the experiment, the recorded EIS spectra evolve. The phase shift at high frequencies increases with prolonged immersion time, whereas the phase shift at low frequencies (5 Hz) generally decreases. Similarly, the impedance curves show an increase at frequencies of approximately 10^2 Hz and a decrease at low frequencies (< 5 Hz). The results of the fitting according to the equivalent circuits shown in Figs. 5.5c and f are presented in Fig. 5.6. The plot of the polarization resistance (R_p) as a function of the immersion time (Fig. 5.6a) reveals clear differences between the alloys. The highest polarization resistance was found for the Fe samples. The TWIP samples show R_p values that are lower by a factor of approximately four. The lowest polarization resistances were found for the TWIP-1Pd samples. It appears that the TWIP-1Pd CW23-700C samples have a higher polarization resistance, but in view of the considerable standard devia-

5.3 Results

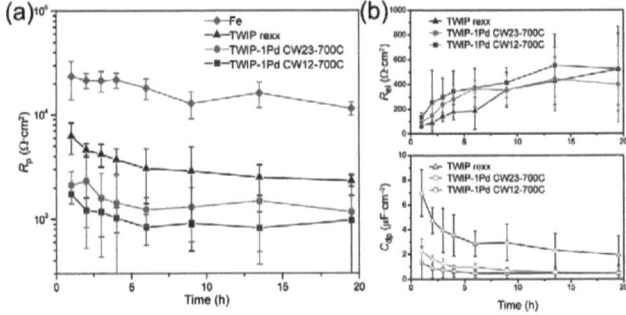

Figure 5.6: Evolution of (a) the polarization resistance, and (b) the electrolyte resistance and the capacitance of the degradation products as a function of the immersion time. The Fe samples feature the highest polarization resistance, which is lower for TWIP and lowest for the TWIP-1Pd samples. The electrolyte resistance values increase for all TWIP(-1Pd) samples with increasing immersion times, representing the evolution of the localized corrosion attacks. Correspondingly, the decrease in the degradation products capacitance values indicates the growth of the degradation product layer on the samples.

tion these differences are not regarded as significant. Common to all samples is a slight decrease in the polarization resistance with prolonged immersion time.

The electrolyte resistance (R_{el}) values determined for the TWIP and TWIP-1Pd samples are plotted in Fig. 5.6b. Their increase with prolonged immersion time indicates the growth of the pits and of the degradation product layers. Even though the R_{el} values are similar for longer (> 10 h) immersion times, it appears that they initially increase more rapidly for the TWIP-1Pd samples. The evolution of the capacitance associated with the constant phase element (Q_{dp}), which represents the degradation products, is also shown in Fig. 5.6b. These values were calculated according to Ref. [35]. It is evident that the TWIP samples feature significantly higher capacitance values than the TWIP-1Pd samples.

The EIS spectra recorded for TWIP-1Pd CW12-700C in SBF(Hepes) (Fig. 5.7) show distinct differences compared to the CO_2-buffered SBF. During the initial 5 h only one time constant is present in the spectra (c.f. the phase angle curves), as previously noted for the Fe samples. The phase angle, however, decreases

5 Degradation properties II - Degradation performance of Fe–Mn–C(–Pd) alloys

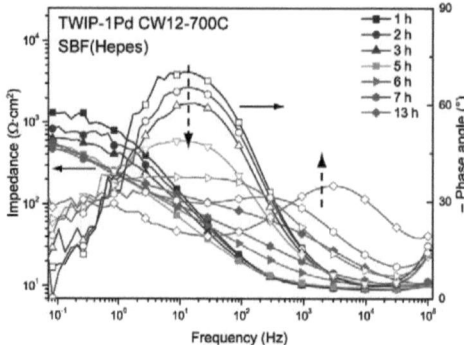

Figure 5.7: Bode representation (impedance and phase angle as a function of frequency) for TWIP-1Pd CW12-700C in Hepes-buffered SBF. Initially, the spectra contain only one time constant. After 7 h of immersion, a second time constant is visible in the spectra, evidence of the pH buffer's influence on the degradation mechanism.

with prolonged immersion time. In the particular case presented in Fig. 5.7, from 7 h onwards a second time constant is present in the spectra. They are then similar to those found previously for the TWIP-1Pd samples (c.f. Fig. 5.5b) in SBF (with CO_2 buffering).

5.3.5 Immersion testing in H_2SO_4

The results from immersion testing in H_2SO_4 (Fig. 5.8) first and foremost reveal the large amount of hydrogen evolution in the TWIP-1Pd samples. In comparison, the hydrogen evolutions of TWIP and Fe are considerably less. This is also apparent from the inset in Fig. 5.8, which shows the hydrogen evolution rate (from a linear fit of the hydrogen evolution) for the different samples. It is particularly interesting to compare the hydrogen evolution rates of the TWIP-1Pd samples at different heat treatment states. Even though there are differences between the individual heat treatment states, a clear trend is hard to identify. It seems that a higher heat treatment temperature causes an increase in the hydrogen evolution rate. However, no clear trend is visible if we compare the samples in the rexx state with those in cold-worked and annealed states. The initial pH-

5.4 Discussion

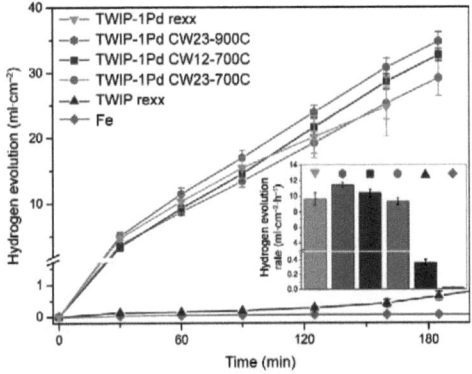

Figure 5.8: Hydrogen evolution data from immersion tests in H_2SO_4. The high degradation rate of the TWIP-1Pd samples is striking. However, there is no clear correlation between microstructure (i.e. precipitate size) and degradation rate (see inset).

value was determined to be 1.4 and remained constant for the Fe and TWIP samples. For the TWIP-1Pd samples the pH-value increased to 1.5 during the immersion period.

5.4 Discussion

In the following, we discuss the results of the immersion testing and EIS experiments, and suggest a degradation mechanism in SBF that also takes into account oxygen reduction as cathodic process. The influence of the precipitates' size and distribution is discussed separately, considering the results of the immersion tests in H_2SO_4.

5.4.1 Degradation behavior in SBF

5.4.1.1 Immersion testing in SBF

Both the EIS results and the immersion tests indicate that the degradation rate can be adjusted according to the principles of the previously described design strategy [11]. The mass loss data from the immersion tests directly reveal the

5 Degradation properties II - Degradation performance of Fe–Mn–C(–Pd) alloys

amount of material that degraded during the tests. Strictly speaking, the mass loss per area displayed in Fig. 5.3 requires a uniform corrosion process [31], which is not a given for all samples. Nevertheless, it is still a useful indicator for establishing a ranking of the different alloys. As previously reported [21], the degradation rate of pure iron (≈ 0.22 mg·cm^{-2}·d^{-1}, corresponding to ≈ 0.10 mm·y^{-1}, determined in the interval from 14 d to 28 d) using the present setup is slightly lower than most literature values (ranging from 0.14 mm·y^{-1} [15] to 0.23 mm·y^{-1} [14], measured in Hank's solution and SBF, respectively). This discrepancy has been attributed to the accelerating effect of pH buffers (such as Tris or Hepes) on degradation. For Mg alloys it has been shown that by using CO_2 to control the pH value of the testing solution the results of in vitro experiments correspond much more to the outcomes of in vivo studies [21].

Alloying with Mn and C (TWIP samples) generates a slightly higher degradation rate of ≈ 0.13 mm·y^{-1}. The accelerating influence of Mn on degradation rate has also been observed by Hermawan et al. [16] for Fe-25Mn (0.52 mm·y^{-1}) and Fe-35Mn (mm·y^{-1}) alloys, and by Liu et al. [17] for Fe-30Mn-6Si (0.30 mm·y^{-1}). Even though the Mn concentrations are roughly in the same range as for the TWIP alloy, the differences are more pronounced in the solutions with pH buffers compared to this study. For Fe-25Mn, the increase in the degradation rate compared to Fe is by a factor of almost 4, whereas in the present study a factor of 1.3 was determined (Fe vs. TWIP). This means that using testing solutions that contain pH buffers may cause the overestimation of in vivo degradation rates. However, this cannot be proven at the moment as there is no in vivo data available for comparison.

The highest degradation rates of ≈ 0.21 mm·y^{-1} and ≈ 0.20 mm·y^{-1} were measured for the TWIP-1Pd CW12-700 and CW23-700 samples, respectively. The increase in the degradation rate compared to the TWIP samples is by a factor of 1.6, which indicates that the addition of only 1 wt.% of Pd is highly effective in increasing the degradation rate.

5.4 Discussion

5.4.1.2 EIS in SBF

The same degradation rates ranking emerged in the EIS results. Polarization resistance decreases with increased alloying content ($R_{p,Fe} > R_{p,TWIP} > R_{p,TWIP-1Pd}$). Because the corrosion current (and therefore also the degradation rate) is inversely proportional to the polarization resistance (Stern–Geary relation [38, 39]), the TWIP-1Pd alloys feature the highest degradation rates. In terms of absolute values, the differences in the R_p values are higher than in the immersion testing results. Impedance spectroscopy is hence a sensitive tool for assessing the degradation properties of degradable implant materials. However, EIS use is not free from pitfalls [31, 40]. Equivalent circuits are necessary to interpret the spectra and to relate the frequency response to electrochemical reactions and processes on the sample and interface. Multiple equivalent circuits often represent an impedance spectrum equally well, then generating different possible values for the individual elements [31]. Information on the degradation mechanism therefore has to be considered when establishing the equivalent circuit used. Diffusion impedance is normally used to reflect the cathodic reaction when investigating Fe corrosion in aerated solutions [34, 38, 40, 41]. However, in the present case it was impossible to accurately fit the impedance spectra via diffusion impedances. This was in part because the time constants of the two processes were close, obscuring the characteristic frequency range in which diffusion behavior is observed, and in part because the high degradation rates interfered with measurements.

The impedance spectra of the Fe samples contain only one time constant and are hence relatively simple. This is not so for the TWIP and TWIP-1Pd samples. Their spectra contain two time constants which have to be appropriately reflected in the corresponding equivalent circuit. As pointed out above, the occurrence of localized corrosion is responsible for the second time constant at high frequencies (approximately 2 kHz) [42]. A similar equivalent circuit was used to model the corrosion mechanism of steel moulds in the pH range between 8 and 11 in contact with a chloride-containing solution [43]. The exact same equivalent circuit as employed in the present study is usually taken to represent the impedance of an electrode coated with an inert porous layer [44]. It was previously reported that (when measuring at the corrosion potential) pits caused by localized corrosion are indistinguishable from pores in an inert porous layer [43, 45, 46]. While the actual geometry of the pores influences the constant phase element C_{dl} [47], R_{el} can be considered as a measure of the length (depth) of the pores / pits [40]. The increasing Rel values shown in Fig. 6b hence indicate the growth of the pits with prolonged immersion time. The rapid initial increase of R_{el} for the TWIP-1Pd samples indicates a higher initial growth rate for the localized attacks than for the TWIP samples. The dielectric capacitance associated with the degradation products (C_{dp}) can be expressed as:

$$C_{dp} = \frac{\epsilon \cdot \epsilon_0}{\delta} \qquad (5.1)$$

where δ represents the film (degradation product) thickness, ϵ the dielectric constant of the material, and ϵ_0 the permittivity of the vacuum [44]. Assuming that ϵ is equal for all samples and constant over the immersion time, δ is inversely proportional to C_{dp}. The data shown in Fig. 5.6b indicate first of all thicker degradation products on the TWIP-1Pd samples compared to TWIP (higher C_{dp} values of the TWIP samples), right from the beginning of the measurements. The C_{dp} values generally decrease with increasing immersion time, indicating the growth of the layer of degradation products.

5.4.2 Degradation mechanism in SBF

In the following, the degradation mechanism of iron in aerated (O_2-containing) neutral solutions is established. It is then taken into account to discuss the findings that the TWIP-1Pd samples have the highest degradation rate and Fe the lowest.

5.4.2.1 Degradation mechanism of Fe-based alloys

The anodic partial reaction (metal dissolution) of Fe and Mn are given in Eqs. 5.2 and 5.3:

$$Fe \rightarrow Fe^{2+} + 2e^- \qquad (5.2)$$

$$Mn \rightarrow Mn^{2+} + 2e^- \qquad (5.3)$$

These reactions proceed rapidly in most media, and when iron corrodes, the rate is usually controlled by the cathodic reaction [48], which is oxygen reduction, Eq. 5.4:

$$H_2O + \frac{1}{2}O_2 + 2e^- \rightarrow 2HO^- \qquad (5.4)$$

Because Fe is the alloys' main constituent, the following gives equations for Fe only. The anodic and cathodic partial reactions need not necessarily take place at the same spot on the surface. Fig. 5.9 shows schematically the process of the degradation (Fig. 5.9a) and formation of degradation products (Fig. 5.9b). The released metal ions react with the hydroxyl-ions to form hydrous ferrous oxide (FeO·nH_2O) or ferrous hydroxide (Fe(OH)$_2$). Eq. 5.5 [16, 48] provides an example:

$$Fe^{2+} + 2HO^- \rightarrow Fe(OH)_2 \qquad (5.5)$$

Because the corrosion reaction proceeds at the metal interface, the layer next to it always consists of FeO·nH_2O or Fe(OH)$_2$. At the outer surface of the hydroxide layer, dissolved oxygen causes the ferrous (Fe^{2+}) oxides to be further converted to hydrous ferric (Fe^{3+}) oxide or ferric hydroxide [48], according to Eq. 5.6:

$$Fe(OH)_2 + \frac{1}{2}H_2O + \frac{1}{4}O_2 \rightarrow Fe(OH)_3 \qquad (5.6)$$

Hydrous ferric oxide is normally orange or brownish (Fe$_2$O$_3$) and is the most visible corrosion product [48]. Degradation products usually have a layered structure, consisting of hydrous Fe$_2$O$_3$ · nH_2O on the top, a black intermediate layer of Fe$_3$O$_4$ · nH_2O, and FeO·nH_2O at the bottom, i.e. at the metal interface [16, 48].

5 Degradation properties II - Degradation performance of Fe–Mn–C(–Pd) alloys

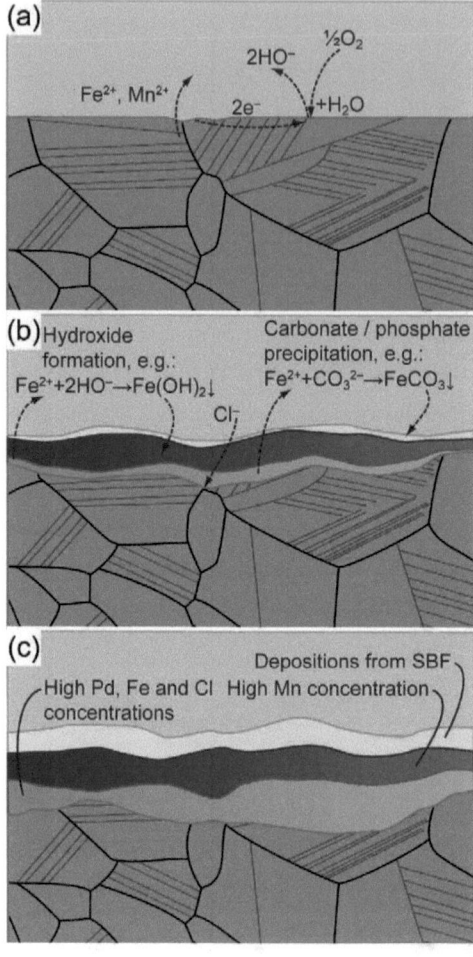

Figure 5.9: Schematic illustration of the degradation process and the formation of degradation products for TWIP-1Pd alloys. (a) Initiation of the corrosion reaction: the anodic partial reactions are the metal oxidation, whereas the cathodic partial reaction is oxygen reduction. (b) Formation of hydroxides / oxides, and precipitation of carbonates and phosphates that comprise the degradation products. Cl^- ions diffuse to the surface, causing localized attacks. (c) Further build-up of degradation products. The Pd remains in the layer close to the sample surface and acts as a macrogalvanic element to increase degradation.

5.4 Discussion

The photographs in Figs. 5.3c-e support the above degradation mechanism. The degradation products appear as red-brown with an underlying black layer which was observed during degradation product removal. Orange-white degradation products are also present on the samples. They correspond to the Ca-rich layers visible in the EDX maps of the cross-sections (c.f. Fig. 5.4). Upon oxidation Fe^{2+} and Mn^{2+} are also able to react with CO_3^{2-} from SBF to form $FeCO_3$ or $MnCO_3$, which have very low solubility constants of log K_{sp} = −11.0 ($FeCO_3$) [49] and log K_{sp} = −11.4 ($MnCO_3$) [50], respectively. However, their precipitation kinetics is reported to be rather slow [50]. On the other hand, the generation of HO− ions from the cathodic reaction (Eq. 5.4) generates a local pH value increase in the vicinity of the sample surface. This may additionally promote the precipitation of carbonates (e.g. $MgCO_3$, particularly influenced by a high HCO_3^- concentration) and phosphates from the SBF [51, 52]. The above compounds eventually generate a decrease in the degradation rate due to mass transport limitations through the precipitates [23, 33, 48, 53, 54]. This study confirms this via the decreasing mass loss rate determined in the immersion tests. Fig. 5.9 also illustrates the precipitation of compounds from the SBF and the layered structure of degradation products that eventually develops (Fig. 5.9c).

The Mn distribution within the degradation products is different than that of Fe (cf. especially Fig. 5.4b). While the highest Fe concentration was found in the layer directly adjacent to the metal surface, Mn concentration was higher in the second and third layers of the degradation products. This may indicate that the Mn ions initially diffuse away from the metal interface and react with the dissolved oxygen at some distance from it to form degradation products.

A high concentration of Cl^- ions was found in the vicinity of the metal surface, indicating that they play an active role in the degradation mechanism. Because the degradation products are porous and do not cover the surface homogenously, Cl^- ions appear to diffuse to the metal surface to compensate the increased cation (Fe^{2+}, Mn^{2+}) concentration (c.f. also Fig. 5.9b). Due to the spatial separation of the anodic and cathodic partial reactions, some metal ions are able to react with Cl^- to form $FeCl_2$, according to Eq. 5.7 [16, 55]:

5 Degradation properties II - Degradation performance of Fe–Mn–C(–Pd) alloys

$$Fe^{2+} + 2Cl^- \rightarrow FeCl_2 \tag{5.7}$$

The metal (Fe, Mn) chloride formed is subsequently hydrolyzed by water, generating free acid (Eq. 5.8) [16] and causing localized corrosion attacks:

$$FeCl_2 + H_2O \rightarrow Fe(OH)_2 + HCl \tag{5.8}$$

5.4.2.2 On the important role of oxygen in the degradation mechanism

As mentioned above (Eq. 5.4), the cathodic partial reaction is oxygen reduction, which is diffusion-controlled, i.e. the oxidation reaction proceeds as rapidly as oxygen reaches the metal surface [48]. In the absence of a diffusion-barrier film on the surface (i.e. degradation products), the theoretical current density is can be calculated according to Eq. 5.9 [48]:

$$i_s = \frac{D \cdot n \cdot F}{d} \cdot c_{O_2} \tag{5.9}$$

where D is the diffusion coefficient for dissolved oxygen in water, $n = 4$ eq·mole^{-1} is the number of electrons transferred, $F = 96500$ C·eq^{-1} is the Faraday constant, d is the thickness of the stagnant layer, and c_{O_2} is the oxygen concentration in the solution.

For the RDE, the corresponding diffusion current density depends on the square root of the rotation rate, as indicated in Eq. 5.10 [46, 56]:

$$i_r = 0.620 \cdot n \cdot F \cdot D^{\frac{2}{3}} \cdot \nu^{-\frac{1}{6}} \cdot c_{O_2} \cdot \omega^{\frac{1}{2}} \tag{5.10}$$

where ν is the kinematic viscosity and ω is the rotation rate.

Consequently, it may be expected that under limited mass-transport conditions (i.e. oxygen reduction) the particular composition of an Fe alloy will have little or no influence on the degradation rate [48]. It is therefore at first glance surprising to note the increases in the degradation rates for TWIP and TWIP-1Pd. However, the above equations (Eqs. 5.9 and 5.10) are only valid for the beginning of the experiments, where the entire metal surface is directly accessible. Thereafter, the formation of degradation products on the samples hinders the oxygen diffusion [46, 48], which generates a decrease in the degradation rate. It is reasonable to assume that the composition and structure of these layers vary for the different alloys. Hence the surfaces are not equally accessible to incoming oxygen.

The EIS results indicate that the polarization resistances, which in fact measure corrosion resistance, decrease from Fe (the highest value) to TWIP-1Pd (the lowest). In addition, they slightly decrease over time, which actually indicates an increasing tendency to corrode. It has been pointed out that the cathodic process (i.e. oxygen reduction) is not entirely mass-controlled but also involve a finite electron transfer [46, 56]. In view of their lower polarization resistance values, this may additionally contribute to the higher degradation rates of TWIP and TWIP-1Pd.

Finally, Pd possesses a high affinity for H_2 and is frequently employed as a catalyst [57, 58]. Pd also accelerates the oxygen reduction reaction [58]; it may be speculated that in the TWIP-1Pd alloys both an additional hydrogen evolution and an increased oxygen reduction rate, catalyzed by the Pd-rich precipitates, contribute to these alloys' enhanced degradation.

5.4.3 Influence of the precipitates on degradation behavior

Pd was deliberately added to the alloys in order to increase their degradation rate. In addition, the original intention was not only to enhance but also to modify the degradation rate by dictating the size and distribution of the Pd-rich precipitates [11]. It was previously reported that precipitates above a critical size cause an increase in the passive current density in Cu-bearing low C steel [59] and an increased tendency for pitting in Al–Cu alloys [60].

Because the differences in the degradation rates are minimal in SBF, immersion tests with the intention to amplify them were performed in H_2SO_4 (Fig. 5.8). If the alloys in the different heat treatment states possess different corrosion1 rates, the relative differences should become more pronounced with increasing overall corrosion rates. Even though the TWIP-1Pd alloys corrode at high rates, it turned out that the size of the precipitates has no apparent influence on the corrosion rates (as indicated in the inset in Fig. 5.8). This in turn means that the mechanical properties (which are determined mainly by the microstructure and the size and distribution of the precipitates [13]) can be optimized without altering the degradation properties.

In view of the Pd distribution in the cross-section of the TWIP-1Pd sample (see Fig. 5.4b), it is interesting to find a high Pd concentration in the layer adjacent to the metal surface. Although the color intensity is relative, it appears that the Pd concentration in this layer is even higher than in the bulk of the material. Significant Pd concentration was also measured in the rest of the degradation products. We therefore conclude that the Pd does not go into solution, but remains incorporated in the degradation products. Since the size of the precipitates is of minor importance we explain the accelerating effect of Pd as follows: the Pd which is deposited on the metal surface acts as a macroscopic, short-circuited galvanic element. Because the Pd-containing layer is in the order of micrometers, small changes in the precipitate's size (in the nm range) are expected to be insignificant.

5.4.4 Influence of degradation products

From the immersion tests (c.f. Figs. 5.3c-e, Fig. 5.4) and RDE experiments it is obvious that degradation product formation actually generates a volume increase. A glaring example of the forces arising from such volume increase is the corrosion-induced failure of steel reinforcements in concrete structures [48, 61]. Although previously noted, the effect of increasing degradation product volumes seems to have been dismissed as unimportant. However, it must be taken into account in assessing these materials for use in degradable medical implants. Undesired increases in degradation product volume may limit the materials' applications, especially in view of the fact that macrophage clearing of degradation products is apparently a slow process [5].

5.4.5 Influence of the testing conditions on degradation behavior

After implantation in a blood vessel, a stent is firstly exposed to blood flow. The in vitro degradation experiment has to mimic this [62]. In both the immersion testing and EIS the solution was agitated for this purpose in the present study. In the EIS measurements the RDE technique also establishes constant hydrodynamic conditions. It is worth noting that the increased flow in the RDE setup apparently favors the occurrence of localized attacks. The (TWIP and TWIP-1Pd) samples show pronounced localized corrosion after 24 h in SBF, but this was not observed to the same extent during the immersion tests (c.f. Figs. 5.3c-d and 5.4), even after the removal of the degradation products. In the RDE setup the degradation products formed during the experiment were partially transported away from the surface, meaning that they can protect it less. The increased oxygen transport in the RDE setup may also have accentuated the differences in the alloys' degradation properties (here expressed by R_p).

It was previously emphasized that using gaseous CO_2 to control the pH value of SBF reflects the in vivo situation more accurately than using a pH buffer [21, 24, 26]. The EIS results in Fig. 5.7 indicate that at first degradation proceeds homogeneously, i.e. only one time constant is present in the spectra as previously reported for Fe. The appearance of a second time constant after 7 h indicates a change in the degradation mechanism, and localized attacks begin. Hence, it is of great importance to realize that deploying a pH buffer directly affects the degradation mechanism in Fe-based alloys.

5.5 Conclusions

In this article the in vitro degradation properties of austenitic TWIP alloys were investigated to evaluate their potential as degradable implant material. To study the plausibility of deployment in temporary cardiovascular stents, their degradation performance was investigated via immersion tests and electrochemical impedance spectroscopy under flow conditions to approximate the in vivo situation. The newly developed TWIP-1Pd alloys reveal a higher degradation rate than pure Fe. This finding can be drawn from both mass loss and electrochemical measurements. EIS proves to be a sensitive tool for investigating the degradation properties of biodegradable Fe-based alloys. Selecting the appropriate equivalent circuit also makes it possible to identify the contributions and evolution of the various active corrosion processes.

The degradation mechanism in SBF was discussed and the decreasing degradation rate (determined from immersion tests) explained by the formation of a layer of degradation products. These consist of hydroxides and carbonates that limit the diffusion of species (oxygen and ions) to the sample surface. Despite the fact that oxygen reduction (a mass-transport-controlled process) is the cathodic reaction, significant differences in the degradation rates can be observed. However, other factors – such as the composition of the degradation products and, possibly electron transfer – also influence degradation behavior and partly explain these findings. It is further seen that the experimental conditions have a substantial influence on the outcome of the experiments. Higher SBF flow, as induced by the RDE experiments, generates an increased tendency for localized attacks, and using a pH buffer clearly affects the degradation mechanism. It is thus of great value to obtain reliable data on in vivo degradation of Fe-based materials to establish an experimental setup capable of predicting their in vivo behavior. As previously performed for Mg-based alloys [63], tests in this context are currently being carried out in a rat model.

Acknowledgements

The authors greatly appreciate financial support received within the framework of the project "Biocompatible Materials and Applications" initiated by the Austrian Institute of Technology GmbH (AIT), and support from the Staub/Kaiser Foundation, Switzerland. The support of the ETH Zurich Electron Microscopy Center is also gratefully acknowledged.

References

[1] Hermawan H, Dubé D, Mantovani D. Developments in metallic biodegradable stents. Acta Biomater 2010;6:1693-7.
[2] Moravej M, Mantovani D. Biodegradable Metals for Cardiovascular Stent Application: Interests and New Opportunities. Int J Mol Sci 2011;12:4250-70.
[3] Zeng R, Dietzel W, Witte F, Hort N, Blawert C. Progress and Challenge for Magnesium Alloys as Biomaterials. Adv Eng Mater 2008;10:B3-B14.
[4] Hänzi AC, Metlar A, Schinhammer M, Aguib H, Lüth TC, Löffler JF, et al. Biodegradable wound-closing devices for gastrointestinal interventions: Degradation performance of the magnesium tip. Mater Sci Eng C 2011;31:1098-103.
[5] Peuster M, Hesse C, Schloo T, Fink C, Beerbaum P, von Schnakenburg C. Long-term biocompatibility of a corrodible peripheral iron stent in the porcine descending aorta. Biomaterials 2006;27:4955-62.
[6] Heublein B, Rohde R, Kaese V, Niemeyer M, Hartung W, Haverich A. Biocorrosion of magnesium alloys: a new principle in cardiovascular implant technology? Heart 2003;89:651-6.
[7] Peuster M, Wohlsein P, Brugmann M, Ehlerding M, Seidler K, Fink C, et al. A novel approach to temporary stenting: degradable cardiovascular stents produced from corrodible metal –results 6-18 months after implantation into New Zealand white rabbits. Heart 2001;86:563-9.
[8] Waksman R, Pakala R, Baffour R, Seabron R, Hellinga D, Tio FO. Short-term effects of biocorrodible iron stents in porcine coronary arteries. J Interv Cardiol 2008;21:15-20.
[9] Hermawan H, Alamdari H, Mantovani D, Dube D. Iron-manganese: new class of metallic degradable biomaterials prepared by powder metallurgy. Powder Metall 2008;51:38-45.
[10] Hermawan H, Dube D, Mantovani D. Degradable metallic biomaterials: Design and development of Fe–Mn alloys for stents. J Biomed Mater Res, Part A 2010;93A:1-11.
[11] Schinhammer M, Hänzi AC, Löffler JF, Uggowitzer PJ. Design strategy for biodegradable Fe-based alloys for medical applications. Acta Biomater 2010;6:1705-13.

5 Degradation properties II - Degradation performance of Fe–Mn–C(–Pd) alloys

[12] Moszner F, Sologubenko AS, Schinhammer M, Lerchbacher C, Hänzi AC, Leitner H, et al. Precipitation hardening of biodegradable Fe–Mn–Pd alloys. Acta Mater 2011;59:981-91.

[13] Schinhammer M, Pecnik CM, Rechberger F, Hänzi AC, Löffler JF, Uggowitzer PJ. Recrystallization behavior, microstructure evolution and mechanical properties of biodegradable Fe–Mn–C(–Pd) TWIP alloys. Acta Mater 2012;60:2746-56.

[14] Zhu S, Huang N, Xu L, Zhang Y, Liu H, Sun H, et al. Biocompatibility of pure iron: In vitro assessment of degradation kinetics and cytotoxicity on endothelial cells. Mater Sci Eng C 2009;29:1589-92.

[15] Moravej M, Purnama A, Fiset M, Couet J, Mantovani D. Electroformed pure iron as a new biomaterial for degradable stents: In vitro degradation and preliminary cell viability studies. Acta Biomater 2010;6:1843-51.

[16] Hermawan H, Purnama A, Dube D, Couet J, Mantovani D. Fe–Mn alloys for metallic biodegradable stents: Degradation and cell viability studies. Acta Biomater 2010;6:1852-60.

[17] Liu B, Zheng YF, Ruan L. In vitro investigation of Fe30Mn6Si shape memory alloy as potential biodegradable metallic material. Mater Lett 2011;65:540-3.

[18] Zhang EL, Chen HY, Shen F. Biocorrosion properties and blood and cell compatibility of pure iron as a biodegradable biomaterial. J Mater Sci: Mater Med 2010;21:2151-63.

[19] Liu B, Zheng YF. Effects of alloying elements (Mn, Co, Al, W, Sn, B, C and S) on biodegradability and in vitro biocompatibility of pure iron. Acta Biomater 2011;7:1407-20.

[20] Nie FL, Zheng YF, Wei SC, Hu C, Yang G. In vitro corrosion, cytotoxicity and hemocompatibility of bulk nanocrystalline pure iron. Biomed Mater 2010;5.

[21] Schinhammer M, Hofstetter J, Wegmann C, Moszner F, Löffler JF, Uggowitzer PJ. On the immersion testing of degradable implant materials in simulated body fluid. submitted to Adv Eng Mater 2012.

[22] Willumeit R, Fischer J, Feyerabend F, Hort N, Bismayer U, Heidrich S, et al. Chemical surface alteration of biodegradable magnesium exposed to corrosion media. Acta Biomater 2011;7:2704-15.

[23] Xin Y, Hu T, Chu PK. In vitro studies of biomedical magnesium alloys in a simulated physiological environment: A review. Acta Biomater 2011;7:1452-9.

[24] Kirkland N, Waterman J, Birbilis N, Dias G, Woodfield T, Hartshorn R, et al. Buffer-regulated biocorrosion of pure magnesium. J Mater Sci: Mater Med 2012;23:283-91.

[25] Kirkland NT, Birbilis N, Walker J, Woodfield T, Dies GJ, Steiger MP. In-vitro dissolution of magnesium-calcium binary alloys: Clarifying the unique role of calcium additions in bioresorbable magnesium implant alloys. J Biomed Mater Res Part B 2010;95B:91-100.

[26] Zainal Abidin NI, Atrens AD, Martin D, Atrens A. Corrosion of high purity Mg, Mg2Zn0.2Mn, ZE41 and AZ91 in Hank's solution at 37 °C. Corros Sci 2011;53:3542-56.

[27] Xin Y, Chu PK. Influence of Tris in simulated body fluid on degradation behavior of pure magnesium. Mater Chem Phys 2010;124:33-5.

[28] Xin Y, Hu T, Chu PK. Degradation behaviour of pure magnesium in simulated body fluids with different concentrations of HCO_3^-. Corros Sci 2011;53:1522-8.

[29] Miller MK. Atom probe tomography: analysis at the atomic level New York: Kluwer Academic / Plenum Publishers 2000.

[30] Müller L, Müller FA. Preparation of SBF with different HCO_3^- content and its influence on the composition of biomimetic apatites. Acta Biomater 2006;2:181-9.

[31] Kirkland NT, Birbilis N, Staiger MP. Assessing the corrosion of biodegradable magnesium implants: A critical review of current methodologies and their limitations. Acta Biomater 2012;8:925-36.

[32] Song GL, Atrens A, StJohn D. An hydrogen evolution method for the estimation of the corrosion rate of magnesium alloys. In: Hryn JN, editor. Magnesium Technology 2001. New Orleans, LA: The Minerals, Metals & Materials Society; 2001. p. 255-62.

[33] Hänzi AC, Gunde P, Schinhammer M, Uggowitzer PJ. On the biodegradation performance of an Mg–Y–RE alloy with various surface conditions in simulated body fluid. Acta Biomater 2009;5:162-71.

[34] Frateur I, Deslouis C, Orazem ME, Tribollet B. Modeling of the cast iron/drinking water system by electrochemical impedance spectroscopy. Electrochim Acta 1999;44:4345-56.

[35] Hsu CH, Mansfeld F. Technical Note: Concerning the Conversion of the Constant Phase Element Parameter Y_0 into a Capacitance. Corros 2001;57:747-8.

[36] de Freitas Cunha Lins V, de Andrade Reis GF, de Araujo CR, Matencio T. Electrochemical impedance spectroscopy and linear polarization applied to evaluation of porosity of phosphate conversion coatings on electrogalvanized steels. Appl Surf Sci 2006;253:2875-84.

[37] De Robertis E, Neves RS, Abrantes LM, Motheo AJ. Pd–P electroless deposition on carbon steel: An electrochemical impedance spectroscopy study. J Electroanal Chem 2005;581:86-92.

[38] Jüttner K. Electrochemical impedance spectroscopy (EIS) of corrosion processes on inhomogeneous surfaces. Electrochim Acta 1990;35:1501-8.

[39] Scully JR. Polarization resistance method for determination of instantaneous corrosion rates. Corros 2000;56:199-218.

[40] Jüttner K, Lorenz WJ, Kendig MW, Mansfeld F. Electrochemical Impedance Spectroscopy on 3-D Inhomogeneous Surfaces. J Electrochem Soc 1988;135:332-9.

[41] Bonnel A, Dabosi F, Deslouis C, Duprat M, Keddam M, Tribollet B. Corrosion Study of a Carbon Steel in Neutral Chloride Solutions by Impedance Techniques. J Electrochem Soc 1983;130:753-61.

[42] Quach NC, Uggowitzer PJ, Schmutz P. Corrosion behaviour of an Mg–Y–RE alloy used in biomedical applications studied by electrochemical techniques. C R Chim 2008;11:1043-54.

[43] Carnot A, Frateur I, Zanna S, Tribollet B, Dubois-Brugger I, Marcus P. Corrosion mechanisms of steel concrete moulds in contact with a demoulding agent studied by EIS and XPS. Corros Sci 2003;45:2513-24.

[44] Orazem ME, Tribollet B. Electrochemical Impedance Spectroscopy. Hoboken, New Jersey: John Wiley & Sons; 2008.

[45] Jovancicevic V, Bockris JOM. The Mechanism of Oxygen Reduction on Iron in Neutral Solutions. J Electrochem Soc 1986;133:1797-807.

[46] You D, Pébère N, Dabosi F. An investigation of the corrosion of pure iron by electrochemical techniques and in situ observations. Corros Sci 1993;34:5-15.

[47] Hitz C, Lasia A. Experimental study and modeling of impedance of the her on porous Ni electrodes. J Electroanal Chem 2001;500:213-22.

[48] Revie WR, Uhlig HH. Corrosion and Corrosion Control. Fourth ed: John Wiley & Sons; 2008.

[49] Bénézeth P, Dandurand JL, Harrichoury JC. Solubility product of siderite ($FeCO_3$) as a function of temperature (25-250 °C). Chem Geol 2009;265:3-12.

[50] Jensen DL, Boddum JK, Tjell JC, Christensen TH. The solubility of rhodochrosite (MnCO$_3$) and siderite (FeCO$_3$) in anaerobic aquatic environments. Appl Geochem 2002;17:503-11.

[51] Yamamoto A, Hiromoto S. Effect of inorganic salts, amino acids and proteins on the degradation of pure magnesium in vitro. Mater Sci Eng C 2009;29:1559-68.

[52] Cornell RM, Schertmann U. The Iron Oxides. Second ed. Weinheim: Wiley-VCH Verlag GmbH & Co.; 2003.

[53] Lindström R, Johansson L-G, Thompson GE, Skeldon P, Svensson J-E. Corrosion of magnesium in humid air. Corros Sci 2004;46:1141-58.

[54] Hänzi AC, Gerber I, Schinhammer M, Löffler JF, Uggowitzer PJ. On the in vitro and in vivo degradation performance and biological response of new biodegradable Mg–Y–Zn alloys. Acta Biomater 2010;6:1824-33.

[55] Yaya K, Khelfaoui Y, Malki B, Kerkar M. Numerical simulations study of the localized corrosion resistance of AISI 316L stainless steel and pure titanium in a simulated body fluid environment. Corros Sci 2011;53:3309-14.

[56] Deslouis C, Gil O, Tribollet B, Vlachos G, Robertson B. Oxygen as a tracer for measurements of steady and turbulent flows. J Appl Electrochem 1992;22:835-42.

[57] Chaplin BP, Reinhard M, Schneider WF, Schüth C, Shapley JR, Strathmann TJ, et al. Critical Review of Pd-Based Catalytic Treatment of Priority Contaminants in Water. Environ Sci Technol 2012;46:3655-70.

[58] Shao M. Palladium-based electrocatalysts for hydrogen oxidation and oxygen reduction reactions. J Power Sources 2011;196:2433-44.

[59] Misawa T, Kobayashi N, Komazaki S, Sugiyama M. Size effect of copper precipitation particles on electrochemical nanoscopic galvanic behavior in Cu-added ultra low carbon steel. Mater Trans 2002;43:1348-51.

[60] Fujimoto S, Iwata T, Tsuji N, Minamino Y. Corrosion behavior of ultra-fine grained Al and Al-2%Cu alloy produced by accumulative roll-bonding (ARB) process. In: Buchheit RG, Kelly RG, Missert NA, Shaw BA, editors. Corrosion and Protection of Light Metal Alloys: Electrochemical Society; 2004. p. 141-7.

[61] Broomfield JP. Corrosion of Steel in Concrete. In: Revie RW, editor. Uhlig's Corrosion Handbook. Hoboken, NJ: John Wiley & Sons; 2011.

[62] Lévesque J, Hermawan H, Dubé D, Mantovani D. Design of a pseudo-physiological test bench specific to the development of biodegradable metallic biomaterials. Acta Biomater 2008;4:284-95.

[63] Kraus T, Fischerauer SF, Hänzi AC, Uggowitzer PJ, Löffler JF, Weinberg AM. Magnesium alloys for temporary implants in osteosynthesis: In vivo studies of their degradation and interaction with bone. Acta Biomater 2012;8:1230-8.

6 Biocompatibility aspects

To evaluate the suitability of Fe-based alloys from a biological point of view, cytocompatibility tests were performed. They are presented in this chapter. Human umbilical vein endothelial cells were used to assess cytocompatibility, based on the alloys' eluates. The results are discussed with respect to the main alloying elements Fe and Mn, and implications for future alloy design are presented.

On the cytocompatibility of biodegradable Fe-based alloys[1]

Biodegradable iron-based alloys are potential candidates for application as temporary implant material. This study summarizes the design strategy applied in the development of biodegradable Fe–Mn–C–Pd alloys and describes the key factors which make them suitable for medical applications. The study's in vitro cytotoxicity tests using human umbilical vein endothelial cells revealed acceptable cytocompatibility based on the alloys' eluates. An analysis of the eluates revealed that Fe is predominantly bound in insoluble degradation products, whereas a considerable amount of Mn is in solution. The investigation's results are discussed using dose-response curves for the main alloying elements Fe and Mn. They show that it is mainly Mn which limits the cytocompatibility of the alloys. The study also supplies a summary of the alloying elements' influence on metabolic processes.

The results and discussion presented are considered important and instructive for future alloy development. The Fe-based alloys developed show an advantageous combination of microstructural, mechanical and biological properties, which makes them interesting as degradable implant material.

6.1 Introduction

Biodegradable metals are considered to be interesting candidates for temporary medical implants, such as cardiovascular stents or for osteosynthesis applications [1-8]. They provide short-term support to the tissue and circumvent long-term clinical issues, e.g. in-stent restenosis, late stent thrombosis or the need to remove the implant in a second clinical intervention [2, 6, 7]. They thus help to improve patient comfort and also reduce the cost of medical treatment.

---

[1]M. Schinhammer, I. Gerber, A.C. Hänzi, P.J. Uggowitzer; Materials Science and Engineering: C 33 (2013) 782-789

6.1 Introduction

In preliminary in vivo studies, stents made of pure iron were investigated [1, 4, 7]. The results were encouraging and showed that iron is a suitable material for degradable stents. In particular, no early restenosis due to thrombotic processes, no pronounced inflammation and no local or systemic toxicity was reported – all in iron's favor [1, 7]. The neointimal proliferation was found to be comparable to that found for standard materials such as stainless steel 316L and cobalt–chromium alloys [4, 7]. However, because iron stents remained mainly intact for up to a year after implantation, a faster degradation rate was targeted [1, 7]. In addition, the mechanical properties of pure iron are modest and not particularly well suited for use as stents. Consequently, research concentrated on these two requirements. Hermawan et al. [9, 10] reported on Fe–Mn alloys which show mechanical properties comparable to stainless steel 316L. Their degradation rates were higher than those of pure iron and adjustable by varying the Mn content [10]. Nevertheless, these degradation rates are still one order of magnitude lower than those of Mg alloys [11] (the alternative engineering metal suitable for degradable implant solutions) and may still be considered too low for practical application. In this context, we developed a design strategy to achieve both a degradation rate and mechanical performance more advantageous than those of pure Fe or Fe–Mn alloys [12]. The austenitic Fe–Mn–C–Pd alloys [13], which are investigated in this study, show especially interesting mechanical performance because they combine the high strength of Co–Cr–Mo alloys (ultimate tensile strength $\approx$ 1200 MPa) with the ductility of stainless steel 316L (uniform elongation $\approx$ 40%). Their superior mechanical performance is ascribed to mechanical twinning during plastic deformation, the so-called TWIP effect (twinning induced plasticity) and the formation of Pd-rich precipitates. These alloys are expected to fulfill the mechanical and electrochemical requirements for stent applications.

In addition to mechanical properties, the biological characteristics of an alloy are important. In vitro cytotoxicity tests are the first step in assessing the biocompatibility of a material. The cytocompatibility of pure iron and iron alloys was previously evaluated using both cell lines (mouse fibroblasts 3T3 and L-929) and primary cells (human umbilical vein endothelial cells (HUVECs) and mouse bone marrow stem cells) [14-20]. Hermawan et al. [14] pointed out that the effect of an alloy on cells may vary solely because of its constituents. It is therefore of interest to investigate the in vitro cytocompatibility of austenitic Fe–Mn–C–Pd alloys. In this study we followed an approach similar to that of Hänzi et al. [21]. We first performed pilot cytocompatibility studies to confirm the suitability of the experimental setup; these served as a basis for the subsequent main cell tests. This approach allows us to compare the results to those published earlier [21]. We also discuss the performance of the Fe-based alloys with respect to their suitability for degradable implants.

6.2 Materials and methods

6.2.1 Materials

Two alloys of nominal composition Fe-21Mn-0.7C (designated as TWIP) and Fe-21Mn-0.7C-1Pd (in wt.%, designated as TWIP-1Pd) in the solution-heat-treated state were used in this study. Their processing, microstructure and mechanical performance are described in detail in [13].

6.2.2 Cytocompatibility studies

NIH 3T3 cells (embryonic mouse fibroblast cell line, European Collection of Cell Cultures) were used in the pilot studies because cell lines are easier to cultivate and more robust than primary cells from tissues. The main cytocompatibility studies were performed using HUVECs (Promocell, Bioconcept, Switzerland), because endothelial cells are in direct contact with the stent surface after implantation.

6.2.2.1 Eluate preparation

Following the protocol established in [21], eluates in simulated body fluid (SBF) were prepared from both alloys. Per alloy, cylindrical samples of 12 mm in diameter and a height of 9.7 mm were ground with 2400 grit SiC abrasive paper and cleaned in ethanol. This ensured that the samples had the same surface area (of 590 mm^2) as previously employed [21]. The samples were then sterilized in 70% ethanol for 5 min, washed three times in SBF, and finally immersed in 10 ml SBF at 37 °C for 24 h and agitated at 180 rpm.

The SBF used in this study is an aqueous solution which mimics the ionic composition and pH value of human blood plasma [21, 22]. Its pH value is buffered using Tris (tris(hydroxymethyl)aminomethane) and was initially adjusted to 7.4 at 37 °C. All solutions deployed for cell testing were sterile-filtered (using filters with a pore size of 0.2 μm) prior to their use.

After immersion, scanning electron microscopy (SEM) images of the corrosion surfaces were obtained using a Hitachi SU-70 (Schottky-type field emission gun) scanning electron microscope at an acceleration voltage of 15 kV. The chemical composition of the surface was approximated using energy dispersive X-ray spectroscopy (EDX) with an Oxford Instruments X-max EDX detector.

6.2.2.2 Pilot and main cytocompatibility studies

For the pilot studies, 3T3 cells were incubated for 5 days in α-MEM supplemented with 10% fetal calf serum and 2 g·ml^{-1} gentamycin at 37 °C in a humidified atmosphere of 5% CO_2 and 95% air. The main studies were performed using HUVECs; the cells were therefore cultured in endothelial cell growth medium containing SupplementMix (ECGM, Promocell, Bioconcept, Switzerland), penicillin and streptomycin. They were inoculated at $\approx 4.5 \cdot 10^3$ cells·cm^{-2} in 48-well plates, and after 24 h the eluates at various volumes ranging from 2 to 200 μl were added to the wells, which contained 500 μl EGCM. This corresponds to 0.4–29% eluate volume per total volume of liquid in the well. The same volumes of SBF and ECGM, respectively, were added to the cells as controls. The cells were incubated for 2 days at 37 °C in a humidified atmosphere of 5% CO_2 and 95% air. Subsequently viability and metabolic activity were determined for both cell types (3T3 and HUVECs). The estimation of viability was based on the physical uptake of neutral red (NR), and metabolic activity was assessed based on an MTT ((3-(4,5-Dimethylthiazol-2-yl)-2,5-diphenyltetrazolium bromide) assay.

The latter is dependent on the activity of intracellular enzymes [23, 24]. Both assays are widely accepted in biocompatibility and cytocompatibility studies [23, 25]. Each measurement point (expressed by mean and standard deviation) is measured in triplicate. The LIVE/DEAD Viability/Cytotoxicity Kit for mammalian cells (Invitrogen) was employed to visualize the live and dead cells by green and red fluorescence, respectively.

6.2.2.3 Dose-response curves

In order to establish dose-response curves for Fe^{2+}, Fe^{3+} and Mn^{2+}, the corresponding chloride salts ($FeCl_2$, $FeCl_3$ and $MnCl_2$) were dissolved in deionized water. The concentration of the cations in the resulting stock solutions was 80 mmol·l^{-1}. Just prior to the tests, the stock solutions were diluted with ECGM to the anticipated final concentration and added to the cells. This procedure was necessary to prevent the precipitation of insoluble Fe- or Mn-compounds from the ECGM, as this would obviously influence the measurement of the optical density. However, at concentrations higher than 10 mmol·l^{-1}, it was difficult to avoid precipitation. Still, the concentration range < 10 mmol·l^{-1} proved to be sufficient for the purpose of this study. The cells (HUVECs) were inoculated at $6.4 \cdot 10^3$ cells·cm^{-2}, and the same assays as described above were employed. ECGM and increasing concentrations of Na^+ (from NaCl) were also added to the cells as control.

6.2.3 Ion release

TWIP and TWIP-1Pd samples (with the same dimensions and surface treatment as described above) were immersed in SBF at 37 °C for 1 week. After immersion for 1, 2, 4 and 7 days the concentrations of Fe and Mn were determined by means of inductively-coupled plasma optical emission spectroscopy (ICP-OES, Paul-Scherrer-Institut, Villigen, Switzerland). Only the concentrations in solution were determined, and potential solid degradation products were not taken into account.

6.3 Results

6.3.1 Pilot cytocompatibility studies

The pilot cell tests using 3T3 cells revealed no adverse influence of the eluates on the cells, even at higher concentrations. In addition, the pilot study also confirmed the suitability of the procedure described in [21], also for biodegradable Fe-based alloys.

6.3.2 Main cytocompatibility studies

Fig. 6.1 shows the influence of the eluates of TWIP and TWIP-1Pd on the viability and metabolic activity of the HUVECs. Significantly, increasing the eluate concentrations by added volumes of 10 µl or higher caused a decrease in both viability and metabolic activity. This indicates that eluate concentrations higher than 2% induced adverse effects in the cells. In contrast, adding SBF only caused a slight decrease in viability and metabolic activity with increasing concentrations. Adding ECGM (as a control) had no adverse effect, but slightly enhanced viability and metabolic activity due to volume increase. Adding eluates from

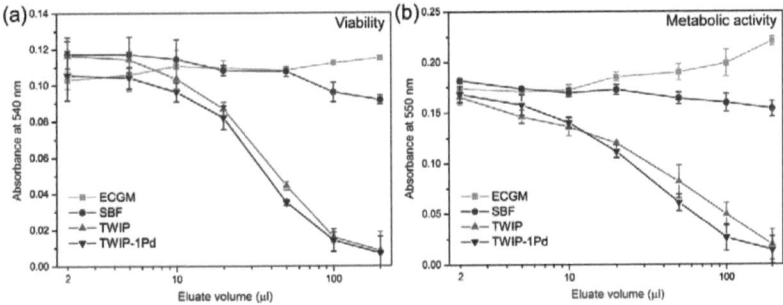

Figure 6.1: Indirect cytocompatibility of HUVECs expressed as: (a) viability (measured by the absorption at 540 nm) and (b) metabolic activity (absorption at 550 nm). The eluates (volumes of 2, 5, 10, 20, 50, 100 and 200 µl) of the alloys TWIP and TWIP-1Pd and the controls SBF and ECGM were added to the 500 µl growth medium in the wells and represent 0.4-29% eluate concentrations.

6 Biocompatibility aspects

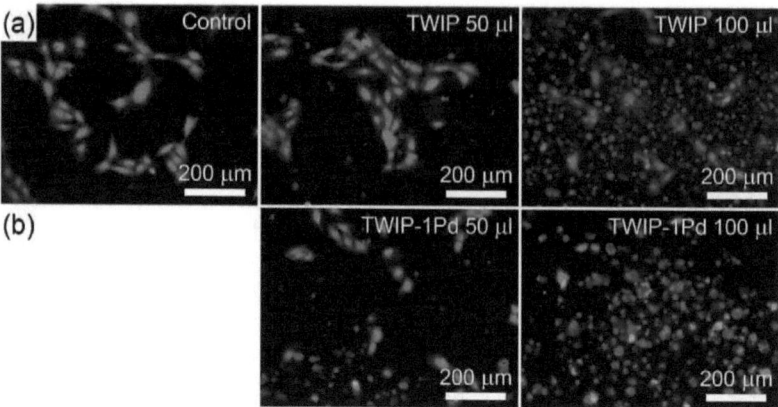

Figure 6.2: The live-dead staining shows the amount of living and dead cells for increasing eluate volumes of the alloys TWIP and TWIP-1Pd. While the addition of 50 μl of eluate generates only a limited number of dead cells, the majority of cells are dead after addition of 100 μl of eluates, irrespective of the alloy's composition.

TWIP and TWIP-1Pd induced similar cell responses. To assess the results' reliability, the experiments were repeated with newly ordered cells and freshly prepared eluates. The findings (data not shown here) confirmed the results shown in Fig. 6.1.

Cells exposed to different volumes of eluates and live-dead stained are depicted in Fig. 6.2. The control image shows only living cells (stained in green), which appeared spread out and securely attached to the well. When 50 μl eluate was added, a considerable number of round, dead cells (stained in red) were found. The ratio of dead to living cells appeared slightly higher for the TWIP-1Pd eluate than for the TWIP eluate. After the addition of 100 μl of eluate most of the cells died.

6.3.3 Dose-response curves

Fig. 6.3 shows the viability and metabolic activity of the HUVECs in response to increasing concentrations of Fe^{2+}, Fe^{3+} and Mn^{2+}. Initially, adding Fe^{2+} ions caused only a slight reduction in viability and metabolic activity. Only

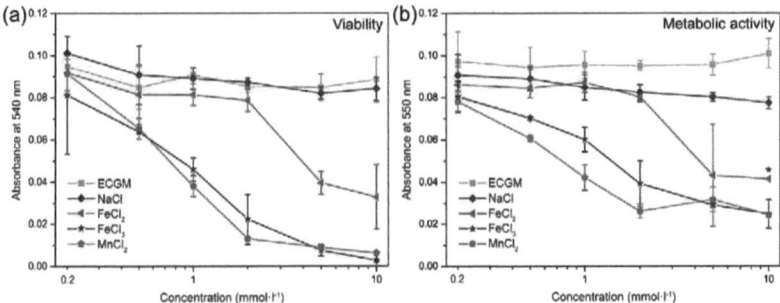

Figure 6.3: Dose-response curves of HUVECs for increasing concentrations of Fe^{2+}, Fe^{3+} and Mn^{2+} ions: (a) viability (measured by the absorption at 540 nm) and (b) metabolic activity (absorption at 550 nm). Increasing concentrations of the $FeCl_2$, $FeCl_3$ and $MnCl_2$ and the controls NaCl and ECGM were added to the 500 µl growth medium in the wells.

* No standard deviation is given because precipitation occurred in all but one well.

concentrations higher than 2 mmol·l^{-1} induced a significant decrease. In contrast, increasing Fe^{3+} and Mn^{2+} ion concentrations generated an immediate decrease in both viability and metabolic activity, i.e. concentrations higher than 0.5 mmol·l^{-1} had an adverse effect on the cells. It appears that the Mn^{2+} ions were tolerated the least, because the descent in viability and metabolic activity is the steepest among the ions investigated here. Adding Na$^+$ ions generated only a slight decrease in metabolic activity, whereas adding ECGM again did not induce negative cell response.

6.3.4 Ion release

Fig. 6.4 shows the Fe and Mn concentrations as a function of immersion time. It is apparent that in SBF only Mn was soluble in considerable concentrations. The Fe concentrations were usually below the blank values of the SBF, whereas the Pd concentrations were below the detection limit (and are therefore not shown). The higher Mn concentrations found for the Pd-containing alloy (TWIP-1Pd) indicate a higher degradation rate than for the TWIP alloy. The parabolic shape of the Mn

6 Biocompatibility aspects

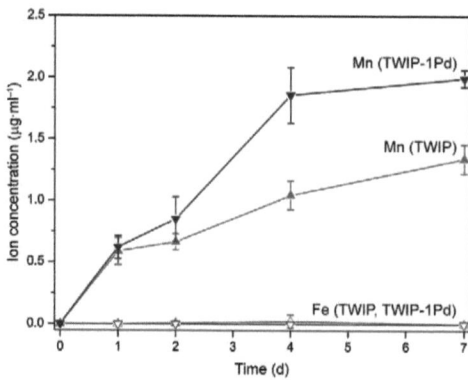

Figure 6.4: Ion concentrations of Fe and Mn after immersion of TWIP and TWIP-1Pd in SBF up to 7 d. The solubility of Fe in SBF is negligible, whereas a significant amount of Mn was measured. The higher Mn concentrations from TWIP-1Pd are due to the higher degradation rate compared to the TWIP samples.

concentration curves derives from the formation of degradation products on the surface that restrict further corrosion and hence decrease the degradation rate [19, 21, 26].

6.3.5 Analysis of sample surface after eluate preparation

The surfaces of the two alloys were similar after eluate preparation, i.e. immersion in SBF for 24 h at 37 °C. In addition to a thin and homogenous degradation layer on the sample surface, spots of localized corrosion were found by visual inspection under the light microscope. They are characterized by an accumulation of red-brown degradation products. The SEM images in Fig. 6.5 show the sample surfaces partly covered with degradation products that appear bright in the secondary electron contrast employed. Apparently the underlying surface was not the bare metal substrate but a uniform layer of degradation products not visible to the naked eye. This is evident from the inset which shows the sample at higher magnification. After removal from the SBF, the samples developed a characteristic "cracked-earth" appearance (marked by the arrow), often encountered after immersion tests [14, 15, 19].

Figure 6.5: Scanning electron images of (a) TWIP and (b) TWIP-1Pd after 24 h in SBF. The surfaces are covered with a homogenous layer of degradation products, which appears dark in the secondary electron contrast. At the top some voluminous and bright degradation products are also seen.

EDX measurements revealed the distinctly different compositions of the areas marked by the squares. The bright and voluminous degradation products contain large amounts of O and are rich in P and Ca. They contain comparatively low amounts of Mn and Fe. In contrast, the underlying degradation products contain hardly any P and Ca, but consist almost solely of O, Fe and Mn. The compositions measured were similar for both alloys, except that for TWIP-1Pd (Fig. 6.5b) Pd was also detected in the underlying degradation products.

6.4 Discussion

Pure Fe [16-19], Fe–Mn- [14, 27] and Fe–Mn–Si-alloys [27], plus the influence of some common alloying elements in steels [15], were investigated in previous cell studies. It was generally found that pure Fe shows acceptable to good cytocompatibility [17-19] and excellent hemocompatibility [17, 18]. The concepts developed to generate a degradation rate quicker than that of pure Fe [9, 12] make use of the accelerating effect of C, Mn and Pd. However, it was found that the alloying elements not only increase the degradation rate of Fe-based alloys, but also decrease their cytocompatibility [14, 15, 27]. This may be caused either by a higher amount of substance release from the samples, or by lower cytocompatibility of the alloying elements.

6.4.1 The alloying elements' influence on metabolic processes and their cytocompatibility

The following sections summarize the role of Fe and Mn in cells and focus on the mechanisms able to provoke toxicity reactions. The use of Pd in biological systems and its cytocompatibility are also discussed below. The aim is to understand the influence of the alloying elements on the cytocompatibility of the alloys under investigation. The dose-response curves presented in Fig. 6.3 assist in the interpretation of the results.

6.4.1.1 Iron cytocompatibility

Iron is essential for fundamental metabolic processes in cells and organisms [28]. Under physiological conditions, extracellular Fe is exclusively bound to transferrin, a monomeric glycoprotein, which keeps Fe soluble and nontoxic [28-30]. This is important for the organism, because under aerobic conditions "free" Fe ions readily catalyze the generation of harmful radicals, i.e. hydroxyl radicals (HO·) from superoxide (O_2^-) and hydroxide peroxide (H_2O_2), collectively termed "reactive oxygen intermediates" (ROIs) [30]. These are formed by means of Fenton and Haber–Weiss chemistry, as indicated in the following. Fenton [31] first observed that a mixture of H_2O_2 and a Fe^{2+} salt produces the HO· radical:

$$Fe^{2+} + H_2O_2 \rightarrow Fe^{3+} + HO\cdot + HO^- \quad (6.1)$$

Superoxide can then reduce Fe^{3+} to molecular oxygen and Fe^{2+}:

$$Fe^{3+} + O_2^- \rightarrow Fe^{2+} + O_2 \quad (6.2)$$

The combination of eq. 6.1 and 6.2 yields the so-called Haber–Weiss reaction [32, 33], in which molecular oxygen, a hydroxyl radical, and a hydroxyl anion are produced from superoxide and hydrogen peroxide. This reaction is catalyzed by trace amounts of Fe:

$$O_2^- + H_2O_2 \xrightarrow{Fe} O_2 + HO\cdot + HO^- \quad (6.3)$$

6.4 Discussion

It is important to note that ROIs (especially O_2^- and H_2O_2) are inevitable byproducts of aerobic respiration. They are generated accidentally, e.g. by incomplete reduction of oxygen in mitochondria or the endoplasmic reticulum [30, 34], or purposefully, e.g. by activated phagocytic cells [29]. Superoxide radicals are formed by the enzyme NADPH oxidase in order to activate defenses against invading pathogens [35]. Redox-active iron may catalyze not only the formation of hydroxyl radicals, but also that of further reactive organic species, e.g. the peroxyl radical (ROO·) [30]. A significant fraction of cellular iron is bound to proteins (e.g. hemoglobin or myoglobin) in the form of heme (a common prostetic group). Heme iron (in both forms, "free" or within hemoproteins) can directly catalyze the formation of radicals [30]. Finally, ferrous iron is able to act as a reactant (rather than a catalyst) in direct interaction with oxygen to form free radicals [30]. The hydroxyl radical in particular reacts readily with almost any molecule in the vicinity of its formation, and causes e.g. oxidation of proteins, peroxidation of membrane lipids, or modification of nucleic acids [17, 30, 35]. The increase in the steady state levels of reactive oxygen species beyond the antioxidant capacity of the organism is termed "oxidative stress". Many pathological conditions, such as chronic inflammation or neurodegeneration, are associated with oxidative stress [30].

Hence the toxicity (in this study expressed by a decrease in viability and metabolic activity at high eluate concentrations) of Fe is usually explained as that higher Fe levels (i.e. more redox-active Fe is available) promote the formation of ROIs [36]. Redox-active Fe is able to readily accept and donate electrons, and can therefore switch easily between Fe^{2+} and Fe^{3+} [8, 16, 30, 35]. In view of eq. 6.1, which states that to form hydroxyl radicals Fe^{2+} is oxidized to Fe^{3+}, it appears counter-intuitive that the critical concentration found for Fe^{3+} is lower than that of Fe^{2+}. However, upon dissolution the chloride salts from Fe react acidically and the pH values of the resulting stock solutions are rather low. They were determined to be 3.2 for $FeCl_2$ and 1.7 for $FeCl_3$. Correction of pH was not possible, as this led to the immediate precipitation of insoluble products. It was previously reported for eluates from Mg-alloys that HUVECs are very sensitive

6 Biocompatibility aspects

to changes in pH value [21], and we therefore assume that the lower tolerance level of Fe^{3+} is caused mainly by the lower pH values of the added solution.

Table 6.1 summarizes the critical concentrations of Fe^{2+}, Fe^{3+}, and Mn^{2+} determined in this study and the literature values from [14, 16, 17, 37-39]. The concentrations are given as IC_{50} values for comparability. The IC_{50} values correspond to the concentrations of substances required to induce a 50% reduction in cellular response in comparison to untreated cells [38]. Generally it can be said that the IC_{50} levels depend strongly on the cell type and testing setup used. This makes comparison of different references difficult, and therefore only general trends and no absolute values can be deduced.

Table 6.1: IC_{50} levels (in $mmol \cdot l^{-1}$) of Fe^{2+}, Fe^{3+}, and Mn^{2+} determined from cytocompatibility tests. The type of cells used is specified for each reference.

Reference	Fe^{2+}	Fe^{3+}	Mn^{2+}	Remarks
This study	≈ 6	1.9	1.1	HUVECs, based on MTT assay
Yamamoto et al. [38]	7.0	5.4	0.05	Murine fibroblasts L929 IC_{50} values correspond to 50% plating efficiency
Contreras et al. [41]	6.8	5	—	3T3 mouse fibroblasts
Sauvant et al. [39]	—	0.77	0.40	Murine fibroblasts L929
Zhang et al. [17]	1.3[a]	—	—	[a] Safe concentration, as stated by the authors, no IC_{50} level given Bone bone marrow cells
Zhu et al. [16]	—	≈ 1.1	—	HUVECs, extracted from figure
Hermawan et al. [14]	—	—	≈ 0.1[b]	[b] Estimated value due to a different setup, 3T3 mouse fibroblasts

In the studies reported by Yamamoto et al. [39] and Contreras et al. [37] on murine fibroblasts, Fe^{2+} ions were tolerated better than Fe^{3+} ions. However, the differences in the IC_{50} levels reported are smaller than those in this study. Zhu et al. [16] also studied HUVECs and determined an IC_{50} level of ≈ 1.1 $mmol \cdot l^{-1}$ for Fe^{3+} ions, which is slightly lower than that of this study. Even though no pH value was given in [16], it is reasonable to assume that the solution used was acidic. It was produced from $NH_4Fe(SO_4)_2$, and upon dissolution both the NH^{4+} and the Fe^{3+} ions reacted acidically, which may also explain the relatively low critical concentration. We therefore conclude that the differences in the IC_{50} levels of Fe^{2+} and Fe^{3+} arise mainly from to the differences in the pH values of the stock solutions, and that the IC_{50} level of Fe can be taken more reliably from the dose-response curve of Fe^{2+}.

6.4.1.2 Manganese cytocompatibility

Manganese is also an element essential for development and body functions, specifically for bone formation, calcium absorption, blood sugar regulation, and fat and carbohydrate metabolism [40, 41]. It forms part of several enzymes, e.g. arginase (which is involved in urea production in the liver), glutamine synthetase (which converts glutamate to glutamine), and superoxide dismutase (which helps to prevent cellular oxidative stress) [41]. In blood plasma most Mn is bound to proteins (Mn^{2+} to albumin and $beta_1$-globulin [41], and Mn^{3+} to transferrin [40]), and only a small amount exists as hydrated ion [40]. Manganese, like Fe, is a redox-active species and can switch between Mn^{2+} and Mn^{3+}. However, although Mn^{2+} is less readily oxidized than Fe^{2+} Mn was shown to be able to induce ROI formation [40, 42, 43]. In addition, due to transport mechanisms that favor influx (via the Ca^{2+} uniporter) over efflux, intracellular Mn^{2+} is sequestered in the mitochondria and these are therefore the primary pool of Mn in the cells [41, 44]. In the mitochondria Mn was shown to inhibit mitochondrial acotinase activity [45], and to disturb the process of oxidative phosphorylation [43, 44]. As already mentioned, physiologically the electron transport chain in mitochondria is an important source of superoxide, and disturbing it provokes an increased concentration of this. Interestingly, Mn serves as an essential cofactor in several mitochondrial enzymes, including superoxide dismutase, which is important in antioxidant cell protection [44, 46]. The role of Mn is hence ambivalent as regards the generation and attenuation of oxidative stress [42, 44]. Finally, Mn is expected to show toxicity via oxidation

6 Biocompatibility aspects

of biological molecules, and by the influencing of Ca and Fe homeostasis [43, 45, 47].

The above arguments explain the limited cytocompatibility found for Mn. In fact, Mn^{2+} ions are tolerated in low concentrations (decrease of viability and metabolic activity at concentrations > 0.5 mmol·l^{-1}, IC$_{50}$ level of 1.1 mmol·l^{-1}) similar to those determined for Fe^{3+}. However, the pH value of the MnCl$_2$-solution was measured at 5.6, and hence is significantly higher to that in the Fe-solutions. Consequently, the low critical concentration of Mn therefore relates to its cytotoxicity and is not an effect of pH. This result agrees with that in studies by Hermawan et al. [14], Yamamoto et al. [39], and Sauvant et al. [38], who all saw either a limited cytocompatibility of Mn^{2+} ions or lower IC50 levels for Mn^{2+} than for Fe^{3+}, respectively.

6.4.1.3 Palladium cytocompatibility

Palladium is a common component of many dental casting alloys [48-50]. Because dental restorations are designed to remain for a long time (10 to 20 years) in the oral cavity, the alloys used for this purpose are designed to be very noble (in terms of electrochemistry) and corrosion-resistant [49, 51, 52]. Nevertheless there is some concern about the slow release of ions from dental restorations and their adverse effects on the body [48, 51, 53]. The biocompatibility of Pd was investigated in several in vitro and in vivo studies (summarized in [48]). The (chemical) form of the metal significantly affects its biological properties. In the case of Pd it was found that it is necessary for Pd to be in its ionic form to cause (adverse) biological effects [48]. The water or lipid solubility of a given compound relate directly to its biological availability and consequently to its toxicity [48]. If a given compound is readily soluble in water or lipid, which enables the metal to pass through the lipid cell membrane, it usually possesses higher toxicity. Among the Pd salts PdCl$_2$ was found to be the least-tolerated compound, whereas metallic Pd appears to cause no adverse reactions [48, 54]. In the cytotoxicity evaluation by Yamamoto et al. [39] Pd (in the form of PdCl$_2$) was found to have cytocompatibility similar to that of Fe. Studies on the ion release from dental alloys where the alloys were immersed in artificial saliva or in cell culture medium showed that no or only very little Pd was released into the solutions [48, 53, 55]. In addition, the concentration of Pd on the sample surface was increased. These results agree with the ICP-OES measurements in this study. The Pd concentrations in the eluates were not measurable, i.e. they were below the detection limit. In a previous study we investigated degradation products after immersion in SBF and found that the Pd is being incorporated in the degradation products [56]. This means that the degrading TWIP-1Pd alloy is not a significant source of Pd ions. In view of the fact that the alloy contains only 0.51 at.%, it is assumed that adding Pd does not limit the cytocompatibility of the alloys under investigation.

6.4.2 Cytocompatibility of TWIP-steels

Because the pilot tests using 3T3 cells indicated no severe adverse effects of the eluates on the cells the main cytocompatibility studies were performed using HUVECs. These cells are more relevant from an application point of view [21] and are also more sensitive than the 3T3 cells. Hence we expected to observe a more pronounced effect of the eluates on the cells. It was found that viability and metabolic activity decrease with increasing eluate concentration. Interestingly, the eluates from both alloys (TWIP and TWIP-1Pd) generated almost the same cell response, even though the degradation rate of TWIP-1Pd was higher than that of TWIP. This was shown by corrosion tests (immersion testing and electrochemical impedance spectroscopy in SBF), which revealed the accelerating influence of Pd on the degradation rate [12]. The eluate volumes where the viability and metabolic activity decreased to 50% of their initial values were found to be 40 and 32 μl for TWIP and TWIP-1Pd, respectively. It appears, though, that the eluates from TWIP are slightly better tolerated.

The ICP-OES measurements revealed that the Fe concentrations in solution are very low. Hence, only a very small amount of the Fe that corroded during the immersion in SBF is in solution, and most precipitated as insoluble degradation products, as also visually observed in the SEM (c.f. Fig. 6.5). In contrast, the concentration of Mn in solution was significantly higher, which means that the solubility of Mn in SBF is much higher than that of Fe. This in turn indicates that the cellular response is mainly determined by the Mn ions in the eluates. The Fe ions are assumed to play only a minor role, as their concentration in the eluates is much lower and because their tolerance limit is significantly higher than that of Mn. The SEM investigations confirmed that the degradation products also contain Fe and Mn.

Because the experimental setup used in this study is identical to that previously employed for to investigate Mg-alloys [21], an approximate comparison with the cytocompatibility of those alloys is possible. Hänzi et al. [21] reported the critical Mg^{2+} ion concentration to be 10 mmol·l^{-1}, thus significantly higher than determined for Fe^{2+} and especially Mn^{2+} ions. For eluates from Mg-alloys, viability and metabolic activity decreased mainly because of volume and pH effects rather than the substances released. Because the pH of the eluates from Fe-based alloys was only slightly higher than 7.4, a pH effect can be ruled out in this study. A cell medium dilution effect due to the addition of SBF to the cells was observed only at higher volumes (approximately 50 μl, c.f. Fig. 6.1) and therefore plays only a minor role. Consequently, the decrease in viability and metabolic activity must be attributed to the substances released from the samples. Regarding the concentrations of degradation products in the eluates, we believe that they are considerably larger than in the in vivo situation due to constant body fluid exchange in the vicinity of the implant. Therefore the study conditions chosen here represent a worst-case scenario.

The cytocompatibility data presented in this study, and previous results [14, 16, 17], indicate that Fe-based alloys are cytocompatible as long as their ion release rate is within tolerable limits. The iron release rate is determined mainly by the in vivo degradation rate, which hence plays a crucial role in the biocompatibility of Fe-based alloys. This also indicates that accelerating the degradation rate as anticipated in in vivo studies [1, 7] must be pursued with the necessary caution, because too high a degradation rate inevitably limits the biocompatibility of Fe-based alloys.

6.4.3 Remarks on experimental setup

Fischer et al. [57] reported that tetrazolium-salt-based assays are influenced by the degradation products of Mg-based alloys. In these assays, metabolically active cells reduce e.g. MTT, and the absorption of the resulting MTT formazan is determined. Fischer et al. found that even in the absence of cells the Mg eluates caused an increase in absorption in the MTT test. This was explained by the ability of Mg^{2+} ions to reduce MTT. To verify the MTT test for the investigation of Fe-based alloys, we also added eluates to wells without cells and determined the absorption. We found no interference, even for high eluate volumes (of 200 μl). The MTT assay is therefore suitable for investigating Fe and Mn cytotoxicity.

In contrast, strong interferences were found for the lactate dehydrogenase (LDH) assay while establishing the dose-response curves. This test measures the enzyme lactate dehydrogenase, which is released in the cell medium upon lysis [58]. It was found that the Fe and Mn ions interact with the chemicals used in the assay, and no reasonable curves could be measured. The LDH assay therefore appears unsuitable for investigating Fe-based alloys.

6.5 Conclusions

In vitro cytotoxicity tests were performed to evaluate the potential of Fe–Mn–C(–Pd) TWIP alloys as biodegradable implant materials. Indirect cell tests using HUVECs were carried out using eluates from the alloys and the results indicated acceptable cytocompatibility. The eluates from TWIP and TWIP-1Pd induced similar cell responses; it was concluded that the alloys perform equally in terms of cytocompatibility. The ICP-OES measurements carried out on the eluates revealed that mainly Mn ions were present from the alloys and that almost all Fe precipitated as insoluble products. The dose-response curves established for Fe^{2+}, Fe^{3+} and Mn^{2+} show a limited cytocompatibility for Mn, which has important implications for future alloy development. It was shown that the pH values of the stock solutions used have a substantial influence on the outcome of dose-response experiments. Therefore this information must be included in the study and taken into account in the interpretation of the results. Because the in vitro model employed represents extreme conditions, we expect the material to possess better in vivo biocompatibility than what appears in in vitro tests, and therefore view it as potentially interesting for use in degradable medical implants.

Acknowledgements

The authors thank H. Sakagami (Meikai University School of Dentistry, Japan) for support with the preparation of the stock solutions and greatly appreciate the financial support received within the framework of the project "Biocompatible Materials and Applications" initiated by the Austrian Institute of Technology GmbH (AIT), and support from the Staub/Kaiser Foundation, Switzerland.

References

[1] Peuster M, Wohlsein P, Brugmann M, Ehlerding M, Seidler K, Fink C, et al. A novel approach to temporary stenting: degradable cardiovascular stents produced from corrodible metal – results 6-18 months after implantation into New Zealand white rabbits. Heart 2001;86:563-9.

[2] Heublein B, Rohde R, Kaese V, Niemeyer M, Hartung W, Haverich A. Biocorrosion of magnesium alloys: a new principle in cardiovascular implant technology? Heart 2003;89:651-6.

[3] Erne P, Schier M, Resink TJ. The road to bioabsorbable stents: Reaching clinical reality? Cardiovasc Intervent Radiol 2006;29:11-6.

[4] Waksman R, Pakala R, Baffour R, Seabron R, Hellinga D, Tio FO. Short-term effects of biocorrodible iron stents in porcine coronary arteries. J Interv Cardiol 2008;21:15-20.

[5] Witte F, Hort N, Vogt C, Cohen S, Kainer KU, Willumeit R, et al. Degradable biomaterials based on magnesium corrosion. Curr Opin Solid State Mater Sci 2009;12:63-72.

[6] Witte F, Kaese V, Haferkamp H, Switzer E, Meyer-Lindenberg A, Wirth CJ, et al. In vivo corrosion of four magnesium alloys and the associated bone response. Biomaterials 2005;26:3557-63.

[7] Peuster M, Hesse C, Schloo T, Fink C, Beerbaum P, von Schnakenburg C. Long-term biocompatibility of a corrodible peripheral iron stent in the porcine descending aorta. Biomaterials 2006;27:4955-62.

[8] Moravej M, Mantovani D. Biodegradable Metals for Cardiovascular Stent Application: Interests and New Opportunities. Int J Mol Sci 2011;12:4250-70.

[9] Hermawan H, Alamdari H, Mantovani D, Dube D. Iron-manganese: new class of metallic degradable biomaterials prepared by powder metallurgy. Powder Metall 2008;51:38-45.

[10] Hermawan H, Dube D, Mantovani D. Degradable metallic biomaterials: Design and development of Fe–Mn alloys for stents. J Biomed Mater Res, Part A 2010;93A:1-11.

[11] Hermawan H, Moravej M, Dubé D, Fiset M, Mantovani D. Degradation Behaviour of Metallic Biomaterials for Degradable Stents. Adv Mater Res 2007;15-17:113-8 [THERMEC 2006 Supplement].

[12] Schinhammer M, Hänzi AC, Löffler JF, Uggowitzer PJ. Design strategy for biodegradable Fe-based alloys for medical applications. Acta Biomater 2010;6:1705-13.

[13] Schinhammer M, Pecnik CM, Rechberger F, Hänzi AC, Löffler JF, Uggowitzer PJ. Recrystallization behavior, microstructure evolution and mechanical properties of biodegradable Fe–Mn–C(–Pd) TWIP alloys. Acta Mater 2012;60:2746-56.

[14] Hermawan H, Purnama A, Dube D, Couet J, Mantovani D. Fe–Mn alloys for metallic biodegradable stents: Degradation and cell viability studies. Acta Biomater 2010;6:1852-60.

[15] Liu B, Zheng YF. Effects of alloying elements (Mn, Co, Al, W, Sn, B, C and S) on biodegradability and in vitro biocompatibility of pure iron. Acta Biomater 2011;7:1407-20.

[16] Zhu S, Huang N, Xu L, Zhang Y, Liu H, Sun H, et al. Biocompatibility of pure iron: In vitro assessment of degradation kinetics and cytotoxicity on endothelial cells. Mater Sci Eng C 2009;29:1589-92.

[17] Zhang EL, Chen HY, Shen F. Biocorrosion properties and blood and cell compatibility of pure iron as a biodegradable biomaterial. J Mater Sci: Mater Med 2010;21:2151-63.

[18] Nie FL, Zheng YF, Wei SC, Hu C, Yang G. In vitro corrosion, cytotoxicity and hemocompatibility of bulk nanocrystalline pure iron. Biomed Mater 2010;5.

[19] Moravej M, Purnama A, Fiset M, Couet J, Mantovani D. Electroformed pure iron as a new biomaterial for degradable stents: In vitro degradation and preliminary cell viability studies. Acta Biomater 2010;6:1843-51.

[20] Mueller PP, May T, Perz A, Hauser H, Peuster M. Control of smooth muscle cell proliferation by ferrous iron. Biomaterials 2006;27:2193-200.

[21] Hänzi AC, Gerber I, Schinhammer M, Löffler JF, Uggowitzer PJ. On the in vitro and in vivo degradation performance and biological response of new biodegradable Mg–Y–Zn alloys. Acta Biomater 2010;6:1824-33.

[22] Müller L, Müller FA. Preparation of SBF with different HCO_3^- content and its influence on the composition of biomimetic apatites. Acta Biomater 2006;2:181-9.

[23] Buzzi S, Jin K, Uggowitzer PJ, Tosatti S, Gerber I, Löffler JF. Cytotoxicity of Zr-based bulk metallic glasses. Intermetallics 2006;14:729-34.

[24] Mosmann T. Rapid colorimetric assay for cellular growth and survival: Application to proliferation and cytotoxicity assays. J Immunol Methods 1983;65:55-63.

[25] Gerber I, ap Gwynn I, Alini M, Wallimann T. Stimulatory effects of creatine on metabolic activity, differentiation and mineralization of primary osteoblast-like cells in monolayer and micromass cell cultures. Eur Cells Mater 2005;10:8-22.

[26] Hänzi AC, Gunde P, Schinhammer M, Uggowitzer PJ. On the biodegradation performance of an Mg–Y–RE alloy with various surface conditions in simulated body fluid. Acta Biomater 2009;5:162-71.

[27] Liu B, Zheng YF, Ruan L. In vitro investigation of Fe30Mn6Si shape memory alloy as potential biodegradable metallic material. Mater Lett 2011;65:540-3.

[28] Hentze MW, Muckenthaler MU, Galy B, Camaschella C. Two to Tango: Regulation of Mammalian Iron Metabolism. Cell 2010;142:24-38.

[29] Halliwell B, Gutteridge JMC. Biologically relevant metal ion-dependent hydroxyl radical generation An update. FEBS Lett 1992;307:108-12.

[30] Papanikolaou G, Pantopoulos K. Iron metabolism and toxicity. Toxicol Appl Pharmacol 2005;202:199-211.

[31] Fenton HJH. Oxidation of tartaric acid in presence of iron. J Chem Soc, Trans 1894;65:899-910.

[32] Haber F, Weiss J. The Catalytic Decomposition of Hydrogen Peroxide by Iron Salts. Proc R Soc Ser A 1934;147:332-51.

[33] Crichton R. Inorganic Biochemistry of Iron Metabolism: From Molecular Mechanisms to Clinical Consequences. Second ed. New York: John Wiley & Sons; 2001.

[34] Halliwell B, Gutteridge JMC. Role of free radicals and catalytic metal ions in human disease: An overview. Methods Enzymol 1990;186:1-85.

[35] Jomova K, Valko M. Advances in metal-induced oxidative stress and human disease. Toxicol 2011;283:65-87.

[36] David A, Lobner D. In vitro cytotoxicity of orthodontic archwires in cortical cell cultures. Eur J Orthod 2004;26:421-6.

[37] Contreras RG, Scougall Vilchis JR, Sakagami H, Nakamura Y, Nakamura Y, Hibino Y, et al. Type of Cell Death Induced by Seven Metals in Cultured Mouse Osteoblastic Cells. In Vivo 2010;24:507-12.

[38] Sauvant MP, Pepin D, Bohatier J, Groliere CA, Guillot J. Toxicity Assessment of 16 Inorganic Environmental Pollutants by Six Bioassays. Ecotoxicol Environ Saf 1997;37:131-40.

[39] Yamamoto A, Honma R, Sumita M. Cytotoxicity evaluation of 43 metal salts using murine fibroblasts and osteoblastic cells. J Biomed Mater Res 1998;39:331-40.

[40] Crossgrove J, Zheng W. Manganese toxicity upon overexposure. NMR Biomed 2004;17:544-53.

[41] Bowman AB, Kwakye GF, Herrero Hernández E, Aschner M. Role of manganese in neurodegenerative diseases. J Trace Elem Med Biol 2011;25:191-203.

[42] HaMai D, Bondy SC. Pro- or anti-oxidant manganese: a suggested mechanism for reconciliation. Neurochem Int 2004;44:223-9.

[43] Reaney SH, Kwik-Uribe CL, Smith DR. Manganese Oxidation State and Its Implications for Toxicity. Chem Res Toxicol 2002;15:1119-26.

[44] HaMai D, Bondy SC. Oxidative Basis of Manganese Neurotoxicity. Ann N Y Acad Sci 2004;1012:129-41.

[45] Chen J-Y, Tsao GC, Zhao Q, Zheng W. Differential Cytotoxicity of Mn(II) and Mn(III): Special Reference to Mitochondrial [Fe-S] Containing Enzymes. Toxicol Appl Pharmacol 2001;175:160-8.

[46] Dobson AW, Erikson KM, Aschner M. Manganese Neurotoxicity. Ann N Y Acad Sci 2004;1012:115-28.

[47] Jynge P, Brurok H, Asplund A, Towart R, Refsum H, Karlsson JOG. Cardiovascular safety of MnDPDP and $MnCl_2$. Acta Radiol 1997;38:740-9.

[48] Wataha JC, Hanks CT. Biological effects of palladium and risk of using palladium in dental casting alloys. J Oral Rehabil 1996;23:309-20.

[49] Kielhorn J, Melber C, Keller D, Mangelsdorf I. Palladium - A review of exposure and effects to human health. Int J Hyg Environ Health 2002;205:417-32.

[50] Geurtsen W. Biocompatibility of dental casting alloys. Crit Rev Oral Biol Med 2002;13:71-84.

[51] Beck KA, Sarantopoulos DM, Kawashima I, Berzins DW. Elemental Release from CoCr and NiCr Alloys Containing Palladium. J Prosthodontics 2012;21:88-93.

[52] Raducanu D, Cimpean A, Vasilescu E, Cojocaru VD, Cinca I, Drob P, et al. Corrosion behaviour and biocompatibility of a new dental noble AuPdAgTi alloy. Mater Corros 2010;61:775-82.

[53] Wataha JC, Nelson SK, Lockwood PE. Elemental release from dental casting alloys into biological media with and without protein. Dent Mater 2001;17:409-14.

[54] Wiester MJ. Cardiovascular Actions of Palladium Compounds in the Unanesthetized Rat. Environ Health Perspect 1975;12:41-4.

[55] Wataha JC, Lockwood PE. Release of elements from dental casting alloys into cell-culture medium over 10 months. Dent Mater 1998;14:158-63.

[56] Schinhammer M, Steiger P, Moszner F, Löffler JF, Uggowitzer PJ. Degradation performance of biodegradable Fe–Mn–C(–Pd) alloys. Submitted to Mater Sci Eng C 2012.

[57] Fischer J, Prosenc MH, Wolff M, Hort N, Willumeit R, Feyerabend F. Interference of magnesium corrosion with tetrazolium-based cytotoxicity assays. Acta Biomater 2010;6:1813-23.

[58] Korzeniewski C, Callewaert DM. An enzyme-release assay for natural cytotoxicity. J Immunol Methods 1983;64:313-20.

7 Summary and Outlook

7 Summary and Outlook

7.1 Summary

The aim of this thesis was to develop and characterize biodegradable Fe-based alloys which are able to fulfill the requirements of temporary medical implants. Generally speaking, there are two main aspects to these requirements: the degradation rate must be appropriate, and strength and ductility must suit the desired application(s). To create Fe-based alloys that meet these requirements, a design strategy was developed. It relies on controlled modification of the Fe matrix via suitable alloying and appropriate thermomechanical treatments. The alloying elements were primarily selected according to the first aspect of requirements: an increase in degradation rate. To this end a large amount of Mn (in the range between 10 and 21 wt.%) was added to make the matrix more susceptible to corrosion. A comparatively small amount of 1 wt.% Pd was also added with the intention of forming noble precipitates which promote active dissolution of the metal. The design strategy and the choice of alloying elements proved very efficient.

The Fe–Mn(–C) system is very interesting from a metallurgical point of view. Depending on the content of Mn and C, a variety of different microstructures (consisting of α- and/or ϵ-martensite and austenite) can be created, which according to composition influence both the active deformation modes and the resulting mechanical performance. While the first alloy investigated (Fe-10Mn-1Pd) is a martensitic high-strength alloy, the austenitic variant (Fe-21Mn-0.7C-1Pd, i.e. TWIP-1Pd) features especially high ductility and pronounced work-hardening capacity. These facts highlight the versatility of the design approach chosen for these alloys. By means of thermomechanical treatments the balance between strength and ductility can be tailored to the intended use. Cardiovascular stent designs, for example, require high ductility and a pronounced strain hardening response (the latter supports the uniform dilation of the stent within the blood vessel). Osteosynthesis requires a high-strength material where somewhat lower ductility is acceptable. High strength makes filigree structures possible, which in turn reduce the burden on the patient. All of these requirements are satisfied by the high-Mn-content alloys investigated in this study. A com-

7.1 Summary

parison with established materials (stainless steel 316L, Ti and its alloys, and Co–Cr–Mo alloys) used in permanent implants today reveals that the TWIP-1Pd alloy combines the high strength of Co–Cr–Mo alloys with the high ductility of 316L, thereby offering more freedom and flexibility in implant design.

Pd, added with the intention to generate Pd-rich precipitates, substantially contributes to the good versatility of TWIP-1Pd. Pd was found to have a very high precipitation tendency in the Fe–Mn(–C) matrix, and the particular system of TWIP steel alloyed with Pd is closely related to microalloyed steels. During annealing treatment of cold-worked samples, both recrystallization and formation of Pd-rich precipitates took place, and it was found that the two processes interact strongly with each other. Precipitate formation is significantly accelerated by the cold-working-induced high defect density in the matrix, which facilitates the rapid diffusion of Pd atoms along dislocation cores and consequently the ready formation of precipitates. The large amount of stored energy is the main driving force for recrystallization. However, because of enrichment with Pd the mobility of dislocations is reduced (solute drag). In addition, the precipitates restrict the grain boundary movement by means of Zener drag. Both effects may strongly impede recovery and recrystallization. By choosing the appropriate parameters, however, recrystallization may even be prevented completely. Carefully chosen parameters were thus employed in the thermomechanical treatment to optimize microstructure and mechanical performance.

In addition to mechanical properties, degradation behavior is a key characteristic of degradable implant materials. Here a reliable correlation between in vitro and in vivo experiments is crucial. This is true not only for Fe but also for Mg alloys. It was found that pH buffers influence degradation behavior because they increase degradation and cause systematic overestimation of the degradation rate. Deploying gaseous CO_2 for pH control makes pH buffers obsolete. Experimental conditions using gaseous CO_2 are hence closer to physiological reality. The pertaining approach described in the corresponding section of this work is modular and can be adapted to various different experimental setups such as immersion tests, electrochemical cells, or laminar flow cells. For the Mg alloy WZ21 the in vitro degradation rate was found to correlate well with results from an in vivo study.

We consequently chose to study the degradation performance of austenitic TWIP alloys using CO_2 pH control. Mass loss data (determined from immersion tests) and electrochemical measurements yielded complementary degradation behavior results. Electrochemical impedance spectroscopy proved a sensitive and valuable tool for identifying the corrosion processes that take place on the sample surface. The TWIP-1Pd alloys feature a higher degradation rate than TWIP alloys and pure Fe. This result supports the considerations of the design strategy. However, the differences in degradation rates between TWIP-1Pd and TWIP, as determined from immersion tests, are lower than predicted by the design strategy and the electrochemical tests. This was attributed to the corrosion mechanism of Fe and its alloys in neutral aerated solutions: oxygen reduction. This process is usually assumed to be diffusion controlled, and thus it is surprising to observe any differences in the degradation rates at all. This anomaly was explained by considering further influences on the degradation rate, such as the structure and composition of degradation products. The latter are also assumed to be responsible for the decreasing degradation rate with prolonged immersion time. However, the full implications of the degradation behavior of the Fe-based alloys for their suitability as implant material have yet to be clarified.

A first step in this direction comprised tests to evaluate the cytocompatibility of the alloys. These tests involved indirect cells (human umbilical vein endothelial cells) and eluates from TWIP(-1Pd) alloys. The results indicate acceptable cytocompatibility. Further insights were gained by establishing dose-response curves for the main alloying elements Fe and Mn. Mn, in particular, was found to have only limited cytocompatibility, which has implications for further alloy development. However, as long as the ion release rate of Mn is not too high, the alloys are expected to be biocompatible. These results naturally require further investigation before the alloy is deployed as implant material.

In summary, the new alloys, especially TWIP-1Pd, feature a promising combination of microstructural, mechanical and electrochemical properties.

7.2 Outlook

The in vivo degradation rate of pure Fe was found to be too low for practical application. Various attempts, including those presented in this thesis, were made to accelerate it. However, in view of the insights on the degradation behavior of Fe-based alloys gained in this study it has yet to be proven that these approaches actually influence in vivo degradation to the extent expected. Unfortunately, at this point there is no alternative to vivo tests that yield quantitative results. These results are required as benchmarks for laboratory experiments and as reliable correlations between in vitro and in vivo results. In vivo findings would facilitate efficient material screening in the laboratory. Results from in vivo studies of Fe and its alloys, however, are scarce, and even those already published contain no quantitative information on the degradation process, such as mass loss as a function of time, loss of cross-section, or degradation product volume.

Several previous studies have shown that Fe-based alloys have a pronounced tendency to form degradation products. From in vitro tests it appears that these products are insoluble and often rather voluminous. It is important to clarify whether the degradation products are potentially harmful, but this phenomenon has never been addressed.

TWIP steels are the focus of current research, not only for their potential as degradable implant material but also for deployment in the automotive industry. Potential applications for these high-ductility alloys include energy absorption elements and improved deep drawing and stretch sheet material forming processes. These applications also exploit the alloys' pronounced strain hardening response. It has been demonstrated that Pd readily forms precipitates in the austenitic matrix if appropriate thermomechanical treatment is applied. So far, only a few attempts have been made to combine TWIP alloys with precipitation strengthening. The insights gained in this study on the interplay between recrystallization and precipitation, however, may also contribute to the development of precipitate-strengthened TWIP steels.

"Alles Wissen und alles Vermehren unseres Wissens endet nicht mit einem Schlusspunkt, sondern mit einem Fragezeichen."

Hermann Hesse

i want morebooks!

Buy your books fast and straightforward online - at one of world's fastest growing online book stores! Environmentally sound due to Print-on-Demand technologies.

Buy your books online at
www.get-morebooks.com

Kaufen Sie Ihre Bücher schnell und unkompliziert online – auf einer der am schnellsten wachsenden Buchhandelsplattformen weltweit! Dank Print-On-Demand umwelt- und ressourcenschonend produziert.

Bücher schneller online kaufen
www.morebooks.de

VDM Verlagsservicegesellschaft mbH
Heinrich-Böcking-Str. 6-8
D - 66121 Saarbrücken

Telefon: +49 681 3720 174
Telefax: +49 681 3720 1749

info@vdm-vsg.de
www.vdm-vsg.de

Printed by Books on Demand GmbH, Norderstedt / Germany